PETROLEUM SCIENCE AND TECHNOLOGY

SEVERE SLUGGING IN OFFSHORE PRODUCTION SYSTEMS

PETROLEUM SCIENCE AND TECHNOLOGY

Severe Slugging in Offshore Production Systems
Saeid Mokhatab
2010. ISBN: 978-1-60876-900-1

PETROLEUM SCIENCE AND TECHNOLOGY

SEVERE SLUGGING IN OFFSHORE PRODUCTION SYSTEMS

SAEID MOKHATAB

Nova Science Publishers, Inc.
New York

LIBRARY OF CONGRESS CATALOGING-IN-PUBLICATION DATA

Severe slugging in offshore production systems / author, Saeid Mokhatab.
    p. cm.
  Includes bibliographical references.
  ISBN 978-1-60876-900-1 (softcover)
  1. Offshore oil well drilling. 2. Petroleum pipelines--Fluid dynamics. 3.
Wells--Fluid dynamics. I. Title.
  TN871.3.M65 2010
  622'.33819--dc22
                            2009054184

Published by Nova Science Publishers, Inc. † *New York*

CONTENTS

Preface   vii

Acknowledgement   ix

Nomenclature   xi

Chapter 1   Introduction   1

Chapter 2   Severe Slugging Characteristics   7

Chapter 3   Modeling of Offshore Production Facilities   35

Chapter 4   Simulation Results and Conclusions   49

Appendix A   77

Appendix B   87

Appendix C   93

Appendix D   97

Index   101

PREFACE

Flexible risers are critical components of floating hydrocarbon production systems since they provide the means of transferring the hydrocarbon from the seabed to the topside structure. A flexible riser can be designed for such systems as different shapes. However, for deep-water depth exceeding 1000 to 1500 m and beyond, the risers are typically designed in free hanging catenary or S-shaped forms. Free hanging configuration is the simplest, and generally the least expensive, riser configuration. Hence, catenary solution has been taken as one of the most important solutions for oil and gas exploitation in deep waters. One important problem experienced in flexible risers is severe slugging phenomenon that typically occurs in the pipeline-riser systems found at offshore platforms. For this system liquid will accumulate in the riser and the pipeline, blocking the flow passage for the gas flow. This results in a compression and pressure build-up in the gas phase that will eventually push the liquid slug up the riser and a large liquid volume be produced into the separator that causes platform trips and plant shut down. The phenomenon is very undesirable due to pressure and flow rate fluctuations, resulting in unwanted flaring and reduces the operating capacity of the separation and compression units. To avoid such problems, it would be desirable to predict the conditions under which severe slugging will occur and be able to change the operating conditions to eliminate it at the design stage and/or to develop methods for suppressing the slugging by control schemes, once slugging appears as a production problem.

In view of the above, the main objective of this manuscript will be quantifying and better understanding of the characteristics of severe slugging both theoretically and experimentally. A review of the existing literature on severe slugging is presented in this manuscript. It also presents the experimental conditions, and describes a new developed model of the offshore production rig, which will be used as a basis to study dynamic interactions between the riser and downstream receiving facilities in the offshore production units. Finally, it details comparing the results of the predictions with experimental data, discussing finding and proposing future avenues for further work.

ACKNOWLEDGEMENT

The experimental work in this manuscript was conducted in the Department of Process and Systems Engineering of the Cranfield University, England, and I owe a great many thanks to those within it. I would like also to convey my appreciation to Peter Wilson and Tzuu Shing Ng at Scandpower, and James Martin and Tracy Wang at AspenTech Inc., for their time and effort in assisting my struggles with OLGA and HYSYS codes, respectively.

NOMENCLATURE

Symbol	Variable	Unit
A	Area	m2
D	Diameter	m
g	Acceleration due to gravity	m/s2
h	Height	m
L	Length	m
m&	Mass Flowrate	Kg/s
MW	Molecular weight	Kg/kmole
NFr	Froude number	------
P	Pressure	Pa or Bar
Q	Volumetric flow rate	m3/hr or lit/sec
R	Ideal gas constant	J/mole. K
T	Temperature	K or C
t	Time	sec
V	Volume	m3 or lit
U	Velocity	m/sec
Z	Compressibility factor	-----
Greek Symbols		
α	Angle of inclination	deg
β	Angle of inclination	deg
ε	Phase fraction or holdup	----
ρ	Density	Kg/m3
λ	No-slip holdup	----
μ	Mass absorption coefficient	m2/Kg
Subscripts		
B	Base	
Bub	Bubble	
D	Drift	
G	Gas	
I	Inlet	
Inj	Injection	

L	Liquid
M	Mixture
O	Outlet
P	Pipeline
R	Riser, Reference
S	Separator or slug tail
TB	Taylor bubble

Superscripts

H	Horizontal
S	Superficial
V	Vertical

Chapter 1

INTRODUCTION

1.1. SUBSEA PRODUCTION

Subsea technology has advanced markedly in the last decades. Much effort has gone into developing suitable hardware, and this is currently continuing as the industry seeks to produce from ever increasing water depths. This has posed additional challenges to the industry in terms of installation of offshore platforms, pipelines and subsea systems. Also, performing such offshore activities means increased cost for offshore developments and thus economic feasibility of the development has to be evaluated. As the oil and gas industry has matured and many existing fields have reached the end of their life, the facilities, which exploited these fields, have to be either decommissioned in an environmentally friendly manner or somehow their lives have to be extended. Extending their lives may allow a reduction in capital expenditure for a new development as the existing oil and gas separation and transportation systems on the matured/expired field can be reused. Also, many new fields, particularly wet gas or high gas-oil ratio (GOR) fields, are not economically viable if the field has to be exploited with its dedicated gas and condensate separation and transportation systems (Amin and Waterson, 2002). However, the economic feasibilities can be met if either the total production of the field is routed to a nearby offshore oil and gas facility, which can handle the additional processing requirements, or it can be transported directly to shore without processing. In both of these cases, total production has to be transported via multiphase flow (gas/liquid hydrocarbons/water) lines. The use of such systems is increasing as it can help achieve economic optimization of the field.

1.2. FLOATING PRODUCTION
IN DEEPWATER LOCATIONS

Floating production systems offer the most practical method for developing medium and deepwater oil and gas reservoirs. Naturally, as exploration and production heads into deeper waters, floating production facilities become increasingly more economical than fixed or guyed-tower and tension-leg platforms (Manning and Thompson, 1995). Floating production systems had been tried on the UK Continental Shelf in the Argyll field in 1975 and used successfully for oil production in other regions, such as Brazil, West Africa, North-western Australia and Southeast Asia. This increased growth is expected to continue into the next decade (Chianis, 2003).

Flexible risers, used to transport production fluids from the seabed to the production vessels, are essential components of floating production, particularly, in harsh environmental conditions where the riser must deal with substantial vessel motion, can be viewed as an established technology in this respect. Figure 1-1 shows some of the different riser shapes developed for these environments (Patel and Seyed, 1995).

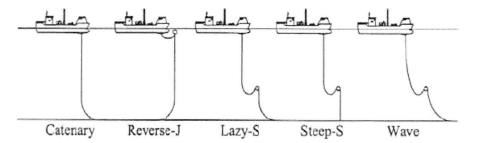

Figure 1.1. Different riser shapes (Patel and Seyed, 1995).

The cost of flexible risers is relatively high and there are technical limitations to the maximum diameter, operating pressure and temperature. Moreover, as water depth increases, the cost of the riser system becomes a higher proportion of the total field development cost. Consequently, it is increasingly important that proper attention is paid to the design and selection of the riser system early in the design loop (Hatton and Howells, 1996).

A considerable part of flexible riser system design is the determination of configuration parameters so that the riser can safely sustain the extreme seastate loadings for which it is to be designed. A well-designed riser configuration is safe and provides compliancy to vessel motions in a cost-effective manner. A riser that is compliant to vessel motions minimizes the station-keeping requirements for the vessel and, in turn, reduces mooring costs.

Figure 1-2 illustrates the generic riser selection process for a deepwater production system. The process starts from knowledge of reservoir and local environmental conditions. The reservoir conditions (requirements or constraints) are the basis for defining the production scheme, which, in turn, is the main input for establishing the subsea layout (definition of the number of subsea manifolds, if any, position of the surface unit, and pipeline route). Conversely, environmental conditions are the main driver for selecting the surface unit concept, where concerning riser interface with the surface unit, two main aspects (motions and riser hang-off) should be considered. However, field layout and surface unit are elements whose definitions affect each other. Moreover, their design attributes affect the riser system, the physical link between them.

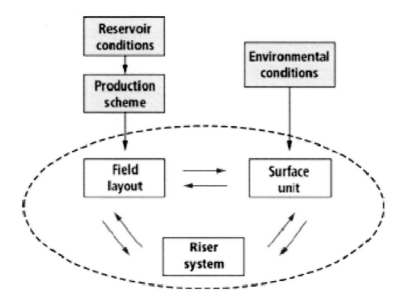

Figure 1.2. Riser system selection process (Sertã, 2004).

In general, the nature of the water current gradients in a particular developments, dictates the riser shape, the type, and relative position of any risers. Free hanging configuration is the simplest, and generally the least expensive, riser configuration. In water depths less than 300 m, catenary riser's application is limited, but in larger water depths the benefits of this alternative riser system can be significant (Hatton and Howells, 1996). Hence, the Steel Catenary Riser (SCR) solution has been taken as one of the most important solutions for oil and gas exploitation in deep waters (Sertã, et al., 1996). A key problem with this solution, however, is that if there are any significant first order wave motions at the vessel connection, the amplitude of dynamic tension is transferred directly to the seabed and this inevitably leads to compression at the riser touchdown point. Furthermore, the free-hanging catenary riser is not very compliant to vessel motions, where riser top tension increases rapidly with far vessel offset, and large vessel offset motions result in correspondingly large and undesirable motions of the riser/seabed touchdown point. The challenge is even more significant when steel catenary risers are considered for harsh environments, where extreme and long-term environmental conditions are amongst the most severe in the world, causing the risers to be highly dynamic and fatigue sensitive (Hatton and Howells, 1996; API RP 17B, 2002; Sertã, 2004).

1.3. MULTIPHASE PRODUCTION PROBLEMS

Multiphase flow transport implies several associated challenges, which can significantly change design requirements (Amin and Waterson, 2002). Hence, one of the objectives of the system designer is to secure "flow assurance", i.e. the transmission system must operate in a safe, efficient and reliable manner throughout design life. Failure to do so has significant economic consequences, in particular for offshore gas production systems. "Flow assurances" is synonymous with wide range of issues that must be constantly addressed throughout the lifetime of a multiphase asset. The term covers the whole range of possible flow problems in pipelines that include both multiphase flow and fluid related effects, e.g. gas hydrate formation, wax deposition on walls, asphaltene precipitation, corrosion, scaling and severe slugging. The avoidance or remediation of these problems is the key aspect of flow assurance that enables the design engineer to optimize the production system and to develop safe and cost-effective operating strategies for the range of expected conditions including start-up, shutdown, and turndown scenarios. However, as production systems go deeper and deeper, flow assurance becomes a major issue for the offshore production systems, where traditional approaches are inappropriate for deepwater production systems due to extreme distances, depths, temperature, or economic constraints (Wilkens, 2002).

1.4. SLUGGING BEHAVIOR
OF PIPELINE/RISER SYSTEMS

Pipeline-riser systems (as shown in Figure 1-3) transport multiphase flow from satellite wells to a central production platform via a single pipeline for economic reasons. In a single multiphase pipeline, however, segregated flow of liquid and gas may cause problems. The actual velocity of the gas phase is faster than the actual liquid velocity. The liquid phase has the tendency to accumulate in the dips and inclined pipe sections causing irregular flow behavior. As a result, large volumes of liquid may flow through the pipeline. These plugs of liquid are called slugs, or riser-induced slugging, and hydrodynamic slugs. Furthermore, operational changes, such as start up and production increase, can create large liquid slugs, severe slugging (Hunt, 1996). For this system liquid slugs will accumulate in the riser and the pipeline, blocking the flow passage for the gas flow. This results in a compression and pressure build-up in the gas phase that will eventually push the liquid slug up the riser and a large liquid volume be produced into the separator, that causes platform trips and plant shut down.

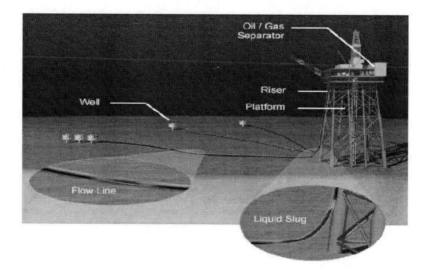

Figure 1.3. Schematic of offshore pipeline-riser system (Henkes, 2004).

The riser slugging phenomenon (also called severe slugging) is very undesirable due to pressure and flow rate fluctuations, resulting in unwanted flaring and reduce the operating capacity of the separation and compression units. To avoid such problems, it would be desirable to predict the conditions under which severe slugging will occur and be able to change the operating conditions to eliminate it in a pipeline-riser system.

1.5. REFERENCES

Amin, R., and Waterson, A., "The Challenges of Multiphase Flow", 14th International Oil & Gas Industry Exhibition and Conference, OSEA 2002, Singapore (Oct./Nov., 2002).

API RP 17B, "Recommended Practice for Flexible Pipe", 3rd Edition, American Petroleum Institute, Washington, USA (March 2002).

Chianis, J.W., "Deepwater Dry-Tree Units for Southeast Asia", *PetroMin,* 29, 8, 30-43 (Oct. 2003).

Hatton, S.A., and Howells, H., "Catenary and Hybrid Risers for Deepwater Locations Worldwide", paper presented at Advances in Riser Technologies Conference, Aberdeen, UK (June 1996).

Henkes, R., "Active Slug Control", paper presented at the 2004 Petronics Workshop, Trondheim (June 15-16, 2004).

Hunt, A., "Fluid Properties Determine Flowline Blockages Potential*", Oil & Gas Journal,* 94, 29, 62-66 (1996).

Manning, F.S., and Thompson, R.E., *"Oil Field Processing,* Vol. 2: Crude Oil", Pennwell Publishing Company, Tulsa, OK (1995).

Patel, M.H., and Seyed, F.B., "Review of Flexible Riser Modeling and Analysis Techniques", *Engineering Structures,* 17, 4, 293-304 (1995).

Sertã, et al., "Steel Catenary Riser for the Marlim Field FPS P-XVIII", paper presented at the Offshore Technology Conference, OTC 8069, Houston, TX (May 6-9, 1996).

Sertã, O.B., "Riser Concepts for Mexican Deepwater Production Systems", *World Oil,* 225, 3 (March 2004).

Wilkens, R.J., Chapter 29: Flow Assurance, in Fluid Flow Handbook, (J.Saleh, Ed.), McGraw-Hill Book Company, New York (2002).

Chapter 2

SEVERE SLUGGING CHARACTERISTICS

2.1. SEVERE SLUGGING IN PIPELINE/RISER SYSTEMS

Severe slugging can occur in multiphase flow systems where a pipeline segment with a downward inclination angle is followed by another segment/riser with an upward inclination angle. For this system, gravity-induced slug flow occurs as a result of a low point connected to an inclining section of the pipe. The pressure drop in the pipeline and the interphase friction between the phases are in these cases not sufficient to transport the liquid uphill in a steady fashion. The liquid will accumulate in the low point, and liquid slug will form. The liquid slug that forms will block the flow of gas in the pipe, and grow until enough upstream pressure has developed to overcome the weight of the liquid slug. These slugs can grow very large, and cause severe problems when they are delivered to the downstream production facility. The inlet separator on the platform will experience large level variation, resulting in poor separation and in some cases flooding. Load variations on the compressors may lead to unnecessary flaring. Another aspect is that the pressure variation caused by slug flow might lead to reduced well performance. To avoid such costly delays, it would be desirable to predict the conditions under which severe slugging will occur and be able to change the operating conditions to eliminate it.

2.1.1. Severe Slugging Mechanism

The process of severe slugging in a pipeline-riser system is considered to consist of four steps; (1) slug formation, (2) slug production, (3) bubble penetration, and (4) gas blow down. This phenomenon had been previously identified by Schmidt et al. (1985) as a cyclic flow rate variation, resulting in periods both of no flow and very high flow rates substantially greater than the time average. Figure 2-1 illustrates the stages of a severe slugging cycle.

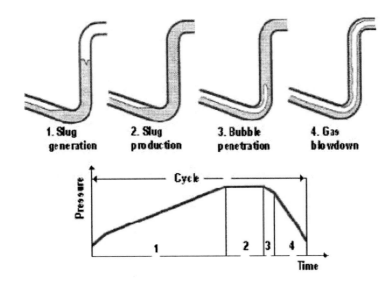

Figure 2.1. Description of severe slugging in pipeline/riser systems (Fabre et al., 1990).

The first step, slug formation, corresponds to an increase of the pressure in bottom of the riser. The liquid level does not reach the top of the riser. During this period, the liquid is no longer supported by the gas and begins to fall, resulting in the riser entrance blockage and the pipeline pressure buildup until the liquid level in the riser reaches to the top. During the second step, slug production, the liquid level reaches the riser outlet, and the liquid slug begins to be produced until the gas reaches the riser base. In third step, bubble penetration, gas is again supplied to the riser, so the hydrostatic pressure decreases. As a result, the gas flow rate increases. The fourth step corresponds to gas blowdown. When the gas produced at the riser bottom reaches the top, the pressure is minimal and the liquid is no longer gas-lifted. The liquid level falls and a new cycle begins (Fabre et al., 1990). This cyclic process becomes steady state when the rate of penetration of the gas into the riser is always positive. However, it is also possible that the penetration of the gas into the riser becomes zero. In this case, liquid blocks the bottom of the riser. This is followed by a movement of the liquid interface into the pipeline and blocking of the gas passage into the riser until the liquid interface reaches the bottom of the riser. At this point, penetration of gas into the riser starts and a new cycle begins again.

When liquid penetrates into the pipeline, the gas in the riser propagates to the top until all of the gas in the riser disappears. When the liquid input is very low, the propagation of the gas towards the top of the riser causes accumulation of all the gas at the top as the liquid falls back. This process is termed cyclic process with fallback, while the former case is termed cyclic process without fallback. In summary, three different possibilities that can occur as a result of penetration of gas into a liquid column in a quasi-steady severe slugging process are identified (Vierkandt, 1988; Jansen et al., 1996):

1. Penetration of the gas that leads to oscillation, ending in a stable steady state flow.
2. Penetration of the gas that leads to a cyclic operation without fall back of liquid.
3. Penetration of the gas that leads to a cyclic operation with fall back of liquid.

The basic process of severe slugging cycle has been also studied and explained by a number of investigators (BØe, 1981; Pots et al., 1985, Fabre et al., 1987; Fuchs, 1987; and Vierkandt, 1988).

Severe slugging in a pipeline-riser system can be considered as a special case of flow in low velocity hilly terrain pipelines, which are often encountered in offshore field. This is a simple case of only one downward inclined section (pipeline), one riser, and constant separator pressure. For this reason, severe slugging has been termed 'terrain-induced slugging' (Fuchs, 1987). This phenomenon has also various names in the industry, including 'riser-base slugging', and 'riser-induced slugging'.

Severe slugging can lead to large surges in the liquid and gas production rates. Figure 2-2 shows some typical predicted time traces accompanying this phenomenon. Clearly such large transient variations will create severe feed disturbances for the topside separator, causing poor separation and in some cases overflow and shutdown of the separator. Oscillations in gas production may cause operational and safety problems during flaring, and the high pressure oscillations can reduce the ultimate recovery from the field, reducing the amount of the recoverable reserves. Other adverse consequences are wear and tear on the equipment resulting in possible unplanned process shutdowns (Yocum, 1973; Schotbot, 1988; Hill, 1990; Sarica and Tengesdal, 2000). Given these potential problems of severe slugging in a pipeline-riser system, the characterization for prevention or control of this phenomenon is highly necessary undertaking.

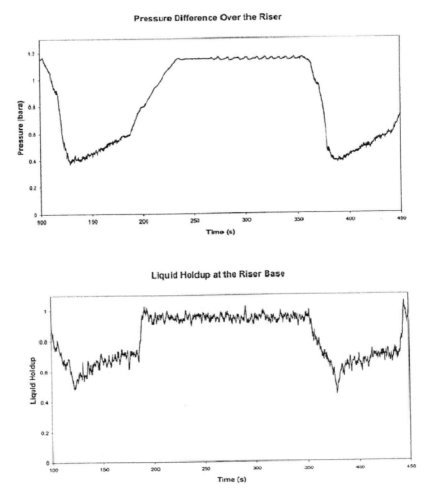

Figure 2.2. Example time traces of surging during severe slugging (Montgomery, 2002).

2.1.2. Flow Pattern Map

The flow characteristics of multiphase flow in a pipeline-riser system are divided into two main regions, steady flow region, and the pressure cycling region, in which the stability line on the flow pattern map separates the two regions. Steady flow includes acceptable slugging, annular-mist and bubble flows. The pressure cycling region includes the severe slugging and transitional regions. In Figure 2-3, Griffith and Wallis (1961) flow map is shown for a pipeline-riser system, which shows regions of stable and unstable behavior. In the above-mentioned figure, N_{Fr} and λ_G are the Froude number of two-phase gas/liquid flow and no-slip gas holdup, respectively given by the following relationships:

$$N_{Fr} = \frac{U_M}{(g\,D)^{0.5}} = \frac{(U_L^S + U_G^S)}{(g\,D)^{0.5}} \tag{2-1}$$

$$\lambda_G = \frac{Q_G}{Q_L + Q_G} \tag{2-2}$$

Yocum (1973) proposed an acceptable transition band based on the field data, where flow to the left of this band was considered acceptable. As can be seen from the figure, at low Froude numbers, bubble flow prevails and fluids will flow through riser pipes without slug formation. However, as Froude number increases the slug flow range is entered.

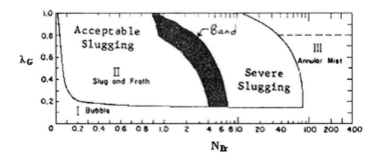

Figure 2.3. Griffith and Wallis flow pattern map with Yocum (1973) transition band (Brill and Beggs, 1991).

Lunde (1989) classified flow in a pipeline-riser system using four main categories; steady flow, oscillation flow, severe slugging, and transitional flow (with occasional severe slugging). Oscillation is a wavy pressure cycling flow phenomenon characterized by a continuous oscillation in liquid and gas production. Several classes of slug flow exist depending upon magnitude of pressure fluctuations and the size and number of alternating gas and liquid slugs.

Tin (1991) described the different types of severe slugging present in an S-shaped riser as following:

1. Severe Slugging 1 (SS1) or Classical severe slugging is characterized with a full column of liquid in the riser prior to gas blow down. The gas blow down is initiated by bubble penetration at the riser base. The liquid slug length is greater than one riser height.
2. Severe Slugging 2 (SS2), where the liquid column is unstable, causing no constant slug production stage. There is no liquid backup along the pipeline; hence the cycle time is generally shorter.
3. Severe Slugging 3 (SS3) is characterized by a continuous bubble penetration in the riser, producing an unstable liquid column in the riser.

Figure 2-4 shows a flow pattern map of the free hanging catenary riser. The same pressure cycling region is observed for different riser configurations and the same stability line is fitted into each flow pattern map. Although the geometry of the riser does not influence the pressure cycling boundary, it does affect the severe slugging boundary and the various cycle characteristics (Tin, 1991).

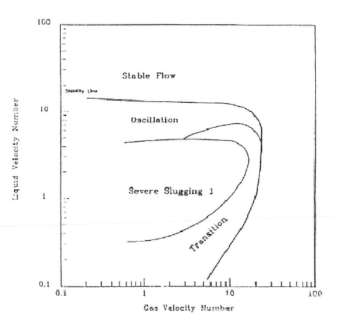

Figure 2.4.Flow pattern map of a free hanging catenary riser (Tin, 1991).

The identification of these flow regimes was based upon visual observation and the pressure profile at the riser base during the slugging process. However, to the design of stable pipeline-riser system, it is necessary to analysis the stability of these systems using computationally intensive parametric techniques. These techniques attempt to build a detailed map of regions of stable and unstable behavior; however, more comprehensive analysis of data (used to build a stability map) is required since considerable uncertainties exist in a number of areas.

2.1.3. Prevention and Control of Severe Slugging

The flow and pressure oscillations due to severe slugging phenomenon have several undesirable effects on the downstream topside facilities unless they are designed to accommodate them. However, designing the topside facilities to accept these transients may dictate large and expensive slug catchers with compression systems equipped with fast responding control systems. This may not be cost-effective and it may be more prudent to design the system to operate in a stable manner (Sarica and Tengesdal, 2000). While slowing production rates can minimize severe slugging, however, operators are looking at alternatives that would allow for maximum production rates without the interruption of slugs (Furlow, 2000). In general, the severe slugging prevention and elimination strategies seek three following approaches.

2.1.3.1. Riser Base Gas Injection

This method provides artificial lift for the liquids, moving them steadily through the riser. This technique can alleviate the problem of severe slugging by changing the flow regime from slug flow to annular or dispersed flow, but does not help with transient slugging in which the liquid column is already formed before it reaches the riser base. It is one of the most used methods for the current applications. However, for deepwater systems increased frictional pressure loss and Joule-Thomson cooling are potential problems resulting from high injection gas flow rates.

Riser base gas injection method was first used to control hydrodynamic slugging in vertical risers. However, Schmidt et al. (1980) dismissed it as not being economically feasible due to the cost of a compressor or pressurizing the gas for injection, and piping required to transport the gas to the base of the riser. Pots et al. (1985) investigated the application of the method to control severe slugging. They concluded that the severity of the cycle was considerably lower for riser injection of about 50% inlet gas flow. Hill (1990) described the riser-base gas injection tests performed in the S.E. Forties field to eliminate severe slugging. The gas injection was shown to reduce the extent of the severe slugging.

The riser base gas lift method may cause additional problems due to Joule-Thomson cooling of the injected gas, where the lift gas will cause cooling and make the flow conditions more susceptible for the wax participation and hydrate formation. Hence, Johal et al. (1997) proposed an alternative technique 'multiphase riser base lift' that requires nearby high capacity multiphase lines diverted to the pipeline-riser system that alleviate the severe slugging problem without exposing the system to other potential problems.

Sarica and Tengesdal (2000) proposed a new technique for sourcing riser base gas lift. The principle of the proposed technique is to connect the riser to the downward inclined segment of the pipeline with a small diameter conduit, where the conduit will transfer the gas from the downward inclined segment to the riser at points above the riser base (multi position gas injection). The transfer process reduces both the hydrostatic head in the riser and the pressure in the pipeline consequently lessening or eliminating the severe slugging occurrence. This method can be considered as self-gas lifting. Sarica and Tengesdal (2000) claimed that the proposed technique is expected to increase production since it does not impose additional backpressure to the production system. The cost of implementing and operation of the proposed systems in the field application is also expected to be low compared to other elimination methods.

Tchambak (2003) conducted a series of experiments on severe slugging mitigation in an S- shaped riser using gas injection. The pipeline/riser system consists of a 2 inch diameter pipe with a 57.4 meters long, 2-degree inclined connected to a 9.982 m high riser. Three injection points were selected for investigation; pipeline, downstream riser base, and base of upper limb. The best injection point was recommended based on the understanding of the gas injection phenomenon and detailed analysis of the data collected. Gas injection reduced the severity of the blow out of liquid and allows a continuous and steady flow of liquid in the pipeline and riser. Gas injection in the pipeline and downstream the riser was reported to effectively eliminating unstable flows in the pipeline-riser. The pipeline injection point was recommended by the author as the best injection for different reasons; lower gas consumption, low liquid peak production, high slug frequency and a steady gas and liquid flow in the riser base at a minimum gas injection flow rate.

2.1.3.2. Topside Choking

This method induces bubble flow or normal slug flow in the riser by increasing the effective backpressure at the riser outlet. While, a topside choke can keep liquids from overwhelming the system, it cannot provide required control of the gas surges that might be difficult for the downstream system to manage. This is a low-cost slug-mitigation option, but its application might be associated with considerable production deferment.

Topside chocking was one of the first methods proposed for the control of severe slugging phenomenon (Yocum, 1973). Yocum observed that increased backpressure could eliminate severe slugging but would severely reduce the flow capacity. Contrary to Yocum's claim, Schmidt et al. (1980) noted that the severe slugging in a pipeline-riser system could be eliminated or minimized by choking at the riser top, causing little or no changes in flow rates and pipeline pressure. Taitel (1986) provided with a theoretical explanation for the success of choking to stabilize the flow as described by Schmidt et al. (1980).

Jansen (1996) investigated different elimination methods such as gas lifting, chocking, and gas lifting and chocking combination. He proposed the stability, and the quasi-equilibrium models for the analysis of the above elimination methods. He experimentally made three observations; (1) large amounts of injected gas were needed to stabilize the flow with gas-lifting technique, (2) careful chocking was needed to stabilize the flow with minimal backpressure increase, (3) gas-lifting and chocking combination were the best elimination method reducing the amount of injected gas and the degree of choking to stabilize the flow.

2.1.3.3. Control Methods

Control methods (feed forward control, slug choking, active feedback control) for slug handling are characterized by the use of process and/or pipeline information to adjust available degrees of freedom (pipeline chokes, pressure and levels) to reduce or eliminate the effect of slugs in the downstream separation and compression unit. Control based strategies are designed based on simulations using rigorous multiphase simulators, process knowledge and iterative procedures. To design efficient control systems, it is therefore advantageous to have an accurate model of the process (Storkaas et al., 2001; Bjune et al., 2002).

The idea of feed-forwarded control is to detect the buildup of slugs and, accordingly, prepare the separators to receive them, e.g. via feed-forwarded control to the separator level and pressure control loops. The aim of slug choking is to avoid overloading the process facilities with liquid or gas. This method makes use of a topside pipeline choke by reducing its opening in the presence of a slug, and thereby protecting the downstream equipment. Like slug choking, active feedback control makes use of a topside choke. However, with dynamic feedback control, the approach is to solve the slug problem by stabilizing the multiphase flow. Using feedback control to prevent severe slugging has been proposed by Hedne and Linga (1990), and by other researchers (Hollenberg et al., 1995; Henriot et al., 1999; Havre et al., 2000; Molyneux et al., 2000; Havre and Dalsmo, 2001; Bjune et al., 2002). The use of feedback control to stabilize an unstable operating point has several advantages. Most importantly, one is able to operate with even, non-oscillatory flow at a pressure drop that would otherwise give severe slugging. Figure 2-5 shows typical applying of active feedback control approach on a production pipeline/riser system, and illustrates how the system uses pressure and temperature measurements (PT & TT) at the pipeline inlet and outlet to adjust the choke valve. If the pipeline flow measurements (FT) are also available, these can be used to adjust the nominal operating point and tuning parameters of the controller.

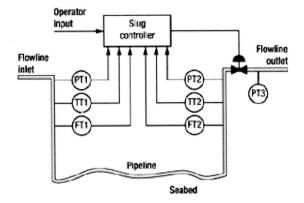

Figure 2.5. Typical configuration of feedback control technique in pipeline/riser systems (Bjune et al., 2002).

Note, the response times of large multiphase chokes are usually too long for such a system to usually be practical. The slug suppression system (S3) developed by Shell has avoided this problem by separating the fluids into a gas and liquid stream, controlling the liquid level in the separator by throttling the liquid stream and controlling the total volumetric flow rate by throttling the gas stream. Hence, the gas control valve back pressures the separator to suppress surges and as it is a gas chock, it is smaller and therefore more responsive than a multiphase chock.

The S3 is a small separator with dynamically controlled valves at the gas and liquid outlets, positioned between the pipeline outlet and the production separator. The outlet's valves are regulated by the control system using signals calculated from locally measured parameters, including pressure and liquid level in the S3 vessel, and gas and liquid flowrates. The objective is to maintain constant total volumetric outflow. The system is designed to suppress severe slugging and decelerate transient slugs so that associated fluids can be produced at controlled rates. In fact, implementation of the S3 results in a stabilized gas and liquid production approximating the ideal production system. Installing S3 is a cost effective modification and has lower capital costs than other slug catchers on production platforms. The slug suppression technology also has two advantages over other slug-mitigation solutions, where unlike a topside choke, the S3 does not cause production deferment and controls gas production, and the S3 controller uses locally measured variables as input variables and is independent of downstream facilities (Kovalev et al., 2003).

2.1.3.4. Other Elimination Techniques

The design of stable pipeline-riser systems is particularly important in deepwater fields, since the propensity towards severe slugging is likely to be greater and the associated surges more pronounced at greater water depths. Therefore, system design and methodology used to control or eliminate severe slugging phenomenon become very crucial when considering the safety of the operation and the limited available space on the platform (Watson et al., 2003). Currently, there are three basic elimination methods that have been already proposed. However, the applicability of current elimination methods to deepwater systems is very much in question (Sarica and Tengesdal, 2000). Anticipating this problem, different techniques have focused to be suitable for different types of problems and production systems.

Yocum (1973) identified different solutions of severe slugging elimination that are still being used today. These are (1) the reduction of the incoming line diameter near the riser to establish a new stable flow regime, (2) install dual or multiple risers, either as separate pipes or as concentric pipes, and (3) the installation of a physical device to remix the fluids in order to prevent the accumulation of liquid at the riser base, and to disturb the flow regime at the riser base and hence prevent the stratified flow regime necessary to cause severe slugging.

One basic method of controlling severe slugging is to prevent the multiphase flow in the pipeline-riser system totally. This is a viable solution that does not impose backpressure on the system. But it requires two separate flow lines and a liquid pump to push the liquids to the surface. The liquid pump is used to overcome the hydrostatic head, therefore preventing the capacity reduction due to backpressure. The stabilization of fluid production to a platform in Malaysia serves as an example of this approach (McGuiness and Cooke, 1993).

2.2. STABILITY ANALYSIS

The stability analysis predicts the boundary between stable (steady flow) and unstable riser flow (severe slugging). Under stable flow conditions no blockage occurs at the bottom of the riser and a steady state flow (bubble or slug flow regime) occurs, while for unstable conditions a liquid blockage and buildup will occur, resulting in a cyclic process. In fact, the results of the stability analysis form a region of a flow regime map, indicating the region of severe slugging. The stability analysis seek to model a particular process required for severe slugging and hence predicts the likelihood of severe slugging, as such these stability models are termed criteria for severe slugging.

The severe slugging process was first modeled by Schmidt et al. (1980). However, their model formed the basis of much of the early works for the stability of severe slugging. Figure 2-6 shows the configuration used for the mathematical criteria presented in this section, which are in common use for the prediction of the occurrence of severe slugging.

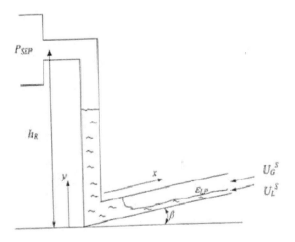

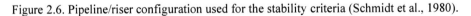

Figure 2.6. Pipeline/riser configuration used for the stability criteria (Schmidt et al., 1980).

2.2.1. Taitel and Dukler (1976) Stratified Flow Criterion

Taitel and Dukler (1976) developed a general criterion for the occurrence of stratified flow in horizontal and near horizontal pipeline. The criterion is based upon considerations of wave growth in a stratified flow (using the inviscid

Kelvin-Helmholtz theory) and gives the maximum allowable superficial gas velocity for stratified flow as:

$$U_G^S < (A_G/A_L) \left[\frac{(\rho_L - \rho_G)g \cos\beta \, A_G}{\rho_G \, dA_L/dh_{LP}} \right]^{0.5} \tag{2-3}$$

Where A is the flow area, β is the angle of inclination, and h is the height of the phase occupying the pipe cross-section. The change in liquid flow area with the liquid height, dA_L/dh_{Lp} is given by Taitel and Dukler (1976) as:

$$\frac{dA_L}{dh_{LP}} = D \sqrt{1 - \left[(\frac{2h_L}{D} - 1) \right]^{0.5}} \tag{2-4}$$

This criterion is used for severe slugging prediction based upon the assertation of Schmidt et al. (1980) that a requirement of "classical severe slugging" in the pipeline-riser system was the occurrence of stratified flow in the pipeline. A typical plot of this criterion is given in Figure 2-7, where the region of stratified flow is the region below the transition line.

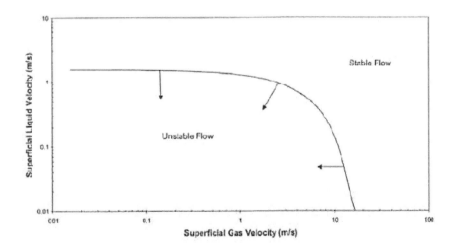

Figure 2.7. Taitel and Dukler (1976) stratified flow criterion.

2.2.2. Boe (1981) Criterion

The Boe (1981) criterion is a simple mathematical expression, which gives the necessary conditions for the occurrence of severe slugging. This criterion has been developed based upon the assertation of Schmidt et al. (1980) that the accumulation of liquid hydrostatic head in the riser must be greater than the pipeline gas-pressure increase, and gives the minimum required liquid velocity for a severe slug to form:

$$U_L^S \geq \left[\frac{P_P}{\rho_L \, g \, (1 - \varepsilon_L) \, L \sin \alpha} \right] U_G^S \tag{2-5}$$

In the above equation, α is the inclination of the riser. Meanwhile, in the initial work by Boe (1981), the no-slip liquid holdup was used, yielded a straight line above which severe slugging occurred (see Figure 2-8). However, other correlations for the liquid holdup yielded an envelope region, where severe slugging would take place.

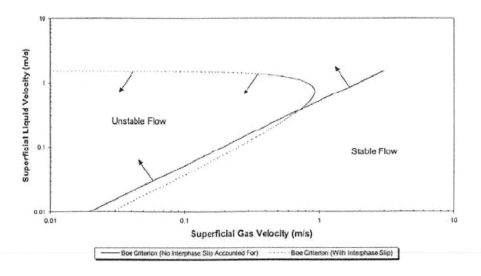

Figure 2.8. Boe (1981) criterion plot.

Note, Equation 2-5 is valid only when no elimination methods are applied (Jansen et al., 1996).

2.2.3. Pots et al. (1985) Criterion

This criterion, termed as Π_{SS} criterion, also considered the formation of a severe slug based upon the rate of the accumulated hydrostatic head in the riser and the pipeline gas-pressure buildup. This analysis leads to the following criterion:

$$\Pi_{SS} = \left(\frac{ZRT/MW}{g\,\varepsilon_L\,L} \right) \left(\frac{\dot{m}_G}{\dot{m}_L} \right) \tag{2-6}$$

In Equation 2-6, Z is the compressibility factor of the gas, MW is the molecular weight of the gas, and $\dot{m}$ is the mass flow rate.

Based on the above criterion, the severe slugging would occur if and only if $\Pi_{SS} < 1$. Pots et al. (1985) also used the value of Π_{SS} to denote the degree of severe slugging, i.e. to quantify the severity of the slugging with a lower Π_{SS}, giving more severe slugs. Figure 2-9 gives a typical plot for the Pots et al. (1985) criterion, with ε_L being calculated using the no-slip condition and the Taitel (1986) correlation for the liquid holdup prediction.

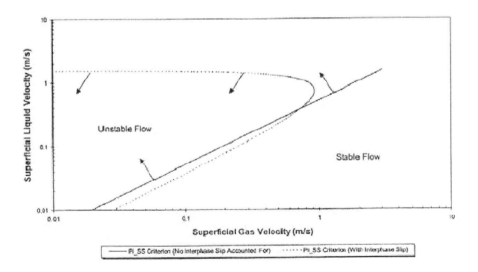

Figure 2.9. Pots et al. (1985) criterion plot.

2.2.4. Taitel (1986) Criterion

This criterion is based on an analysis of the propagation of a bubble penetrating the riser base, causing the liquid column in the riser to become unstable and precipitating blowdown of the riser. The analysis leads to the following criterion in terms of the pipeline gas holdup and the pressure in the separator:

$$\frac{P_S}{P_o} < \frac{(\varepsilon_{GP} / \varepsilon_G') L - h_R}{P_0/\rho_L \, g} \qquad (2\text{-}7)$$

Where ε_G' is the gas holdup immediately behind the penetrating Taylor bubble front. Taitel (1986) assumed that the gas holdup in the bubble tail is a constant (0.89) for vertical flow. Therefore, in order to plot the results on a flow regime map, an expression for the pipeline gas holdup (ε_{GP}) is required. As only a single value of ε_{GP} is obtained for any given separator pressure and the pipeline liquid holdup is a function of the superficial liquid velocity that can be derived from the gas fraction, the criterion gives a limiting liquid velocity for severe slugging (see Figure 2-10).

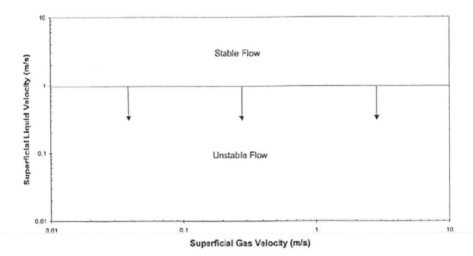

Figure 2.10. Taitel (1986) criterion plot.

2.2.5. Fuchs (1987) Pressure Criterion

This criterion has been developed based upon consideration of this point that during the release of a severe slug, the tail of the severe slug will accelerate through the riser. This release is the moment in time where the slug tail moves from the pipeline into the riser, proper leading to gas blow down. This gives the basic form of the Fuchs (1987) criterion for the release of a severe slug, i.e. the initiation of bubble penetration:

$$\frac{d(P_P - P_S)}{dt} > \frac{d(\Delta P_{Hyd})}{dt} \tag{2-8}$$

Splitting the left hand side of the above equation gives:

$$\frac{dP_P}{dt} > \frac{d(P_S + \Delta P_{Hyd})}{dt} \tag{2-9}$$

Where at the start of bubble penetration stage, the right hand side of Equation 2-9 corresponds to the riser base pressure, therefore, the basic form of Fuchs (1987) criterion for the release of a severe slug can be written as:

$$\frac{dP_P}{dt} > \frac{dP_B}{dt} \tag{2-10}$$

The pressure changes in the pipeline gas and at the riser base are obtained from a mass balance on the gas in the pipeline and a pressure balance over the riser, respectively. Resolving Equation 2-10 in terms of the gas velocity entering the riser base, and using the ideal gas equation of state, gives the final form of the criterion as:

$$\frac{\left(\frac{P_S A}{V_{GP}}\right) + \left(\frac{P_P A}{V_{GS}}\right)}{g \sin\beta (\rho_L - \rho_G)} < \frac{\varepsilon_L - \varepsilon_L'}{1 - \varepsilon_L'} \left(\frac{U_{GB}^S}{\left(U_{GB}^S + U_{LB}^S\right) - \left(U_{GI}^S + U_{LI}^S\right)} \right) \tag{2-11}$$

Where in the above equation, the subscripts GB and LB refer to the gas and liquid entering at the riser base.

Note, although during severe slugging the bubbles continue to accelerate continuously through the riser length, however, the acceleration of bubbles at the riser base is not unique to severe slugging, where during hydrodynamic slug and bubble flow, gas bubbles accelerate during the first instance of bubble penetration to their steady velocity further up the riser.

2.2.6. Jansen et al. (1996) Criterion

This analysis utilizes the stability concept presented by Taitel (1986) and performs an overall force balance including the effects of the choke and gas lift. For the case of choking, Jansen et al. (1996) carried out a theoretical and experimental investigation of choking, modifying the Taitel (1986) criterion to account for the additional backpressure of choking, so that the criterion for severe slugging became:

$$\frac{P_S + C U_L^{S\,2}}{P_o} > \frac{\dfrac{\varepsilon_G}{\varepsilon_G'} L \left(1 - \dfrac{K}{\rho_L g}\right) - h_R}{\dfrac{P_o}{\rho_L g}} \tag{2-12}$$

Where C is the choke coefficient that can be determined experimentally, and K is a proportionality constant.

If this criterion is satisfied, the introducing bubble will not be accelerated up the riser and no blowout will occur. This will lead to a stable flow in the riser.

Jansen et al. (1996) also extended the Taitel (1986) stability criterion for severe slugging to include steady-state gas injection into the riser base. For the case where only gas from the gas lift flows in the riser and no pipeline gas penetrates the riser, the stability criterion is given as:

$$\frac{P_S}{P_0} > \frac{\dfrac{\varepsilon_{GP}}{\varepsilon_G'} L - h_R}{\dfrac{P_o}{\varepsilon_{GR} \rho_L g}} \tag{2-13}$$

where:

$$\varepsilon_{GR} = 1 - \frac{U_{GR}^S}{U_{Bub}} \qquad\qquad (2\text{-}13\text{-}1)$$

$$U_{Bub} = C_0 U_M + U_D \qquad\qquad (2\text{-}13\text{-}2)$$

Where C_0 and U_D are the drift parameter and the bubble drift velocity, respectively. For fully developed Taylor bubbles the values are $C_0 = 1.2$ and $U_D = 0.35 (g\ D)^{0.5}$. For bubble flow the values are $C_0 = 1.0$ and U_D is given by Harmathy (1960) equation.

The Jansen et al. (1996) criteria were tested against severe slugging experimental data (collected in a 25.4 mm diameter, 9.1 m long pipeline, connected to a 3 m high riser that can be inclined from $+5^o$ to -5° from the horizontal) and a successful prediction of the severe slugging region by the criteria was demonstrated.

2.2.7. Montgomery (2002) Criterion

This criterion has been developed for the occurrence of severe slugging in an S-shaped riser, based upon considerations of bubble penetration at the riser base, so that for severe slugging to form it was required that there was no gas penetration at the riser base prior to the filling of the upward limbs of the riser, to be blown out by the pipeline gas during the cycle. The criterion states that in order to prevent a bubble penetrating the riser base, the inlet gas superficial velocity must be lower than the superficial velocity of gas entering the riser base, given by:

$$U_{GI}^S \leq \left[1 - \frac{g\ \varepsilon_{LP}\ L_P\ \sin\beta\ (\rho_L - \rho_G)}{P_S + \rho_L\ g\ h_R} \right] U_{GB}^S \qquad\qquad (2\text{-}14)$$

Where the superficial velocity of gas entering the riser base, U_{GB}^S, can be obtained as:

$$U_{GB}^S = \left(C_0\ U_{LI}^S + U_D \right) \left(\frac{\varepsilon_G'}{1 - \varepsilon_G'\ C_0} \right) \qquad\qquad (2\text{-}15)$$

Note, one assumption is made in this analysis that the gas fraction behind the bubble nose would be less than or equal to the pipeline gas fraction. In fact:

$$\varepsilon'_G = \min\left(\varepsilon_{Gp}, \varepsilon_{GTB}\right) \qquad (2\text{-}16)$$

Where the gas fraction behind a Taylor bubble, ε_{GTB}, can be determined by the Carew et al. (1995) correlation as:

$$\varepsilon_{GTB} = 0.59 + 0.3031\left(\frac{\beta}{90}\right)^{0.2308} \qquad (2\text{-}17)$$

The distribution parameter, C_0, and the drift velocity, U_D, are both functions of pipe inclination and the flow velocities that can be calculated by Bendiksen (1984) relations as:

$$C_0(\beta) = \begin{cases} C_0^H + (C_0^V - C_0^H)\sin^2\beta & \text{if } N_{Fr,C} < 3.5 \\ 1.2 & \text{if } N_{Fr,C} > 3.5 \end{cases} \qquad (2\text{-}18)$$

Where C_0^H and C_0^V are the horizontal and vertical values of the distribution parameter, respectively, with values of 1.05 and 1.2 respectively. The critical Froude number, $N_{Fr,C}$ is also calculated as:

$$N_{Fr,C} = \frac{U_L^S}{\sqrt{g.D_P}} \qquad (2\text{-}19)$$

Also:

$$U_D = \left(0.54\sqrt{g\,D_P}\right)\cos\beta + \left(0.35\sqrt{g\,D_P}\right)\sin\beta \qquad (2\text{-}20)$$

The criterion results were compared against experimental data and showed that the region of severe slugging 1 and pressure cycling are encompassed by the criterion, though there is some over prediction in terms of the maximum superficial gas velocity for unstable flows. Note, comparing to existing criteria, Montgomery (2002) criterion showed improved performance in terms of the limiting conditions for unstable flows. However, the generality of this criterion remains to be confirmed, particularly in different riser shapes.

2.2.8. Tchambak (2004) Criteria

Tchambak (2004) presented the flow stability criteria predicting regions of unstable and stable flows in an S-shaped riser for gas injection near the pipeline inlet, upstream and downstream the riser base. The criteria, as presented below, were successfully validated with different experimental data.

Stability criterion for gas injection near the pipeline inlet:

$$U_{G,Inj}^S + U_{GI}^S > U_{GB}^S \left[1 - \frac{g\,(\rho_L - \rho_G)\sin\beta\,\varepsilon_{GP}\,L}{P_P} \right] \tag{2-21}$$

Stability criterion for gas injection downstream the riser base:

$$U_{GI}^S > U_{GB}^S \left[1 + \frac{\varepsilon_{GP}\,g\,L\left(\rho_L\,\dfrac{\varepsilon_{GR} - \varepsilon_{GS}}{\varepsilon_{GS}} + \rho_G\right)\sin\beta}{P_P} \right] \tag{2-22}$$

Stability criterion for gas injection upstream the riser base:

$$U_{G,Inj}^S + U_{GI}^S > U_{GB}^S \left[1 + \frac{g\left(\rho_L\,\dfrac{\varepsilon_{GR} - \varepsilon_{GS}}{\varepsilon_{GS}} + \rho_G\right)(\varepsilon_{GP}A_P\,L + U_{Bub})\sin\beta}{A_P\,P_P} \right] + \frac{A_{Bub}U_{Bub}}{A_P} \tag{2-23}$$

The criteria demonstrated successful prediction of the severe slugging region when gas injection is used. Each model was individually fitted to a relevant experimental data and the criterion boundary was able to delineate the region of stable and unstable flows with a high degree of accuracy. With the knowledge gained from experiments, the author claimed that injecting gas near the pipeline inlet was similar as if the total gas was inputted as inlet flow at the pipeline inlet (i.e. as if no gas injection was applied). Therefore, the stability criterion of gas injection near the pipeline inlet was compared with data collected by Tin (1991) and Montgomery (2002), and the ability of the criterion to adequately outline the two boundaries was evidenced.

2.3. Severe Slugging in Flexible Risers

This section tries to present the research studies that have been carried out on severe slugging in flexible risers in two parts: (1) experiments on different riser configurations, and (2) existing transient code predictions.

2.3.1. Experimental Investigations

The first reported experimental study of severe slugging in a flexible riser has been presented by Tin (1991). Experimental studies were performed with a two-phase mixture of air and water on three different configurations which included free hanging catenary, lazy S, and steep S riser shapes. The simulated pipeline/riser system consists of a 2 inch diameter pipe with a 60 meters long, 2-degree inclined pipeline terminated in a 33 meters high riser. The effects of changing riser geometry on severe slugging characteristics were identified and accurate flow pattern maps for different riser configurations have been produced. In addition, within the pressure cycling region further detailed experiments were performed to establish pressure cycling characteristics in the different severe slugging regions. From analysis of the flow behavior, it was found that the pressure cycling behavior of an S shaped risers varies significantly from that found in vertical risers, and can be considered as two free hanging catenary risers connected together with each riser having independent effects on the cycle. Within the current experimental accuracy, the lazy S riser can be seen to be more advantages than a free hanging catenary riser when operated within the pressure cycling region.

Different flow regime maps were also identified by Tin and Sarshar (1993), showing the limits of the severe slugging and unstable flow regions. These maps showed that for the different riser geometry, the region of severe slugging and unstable flows was the same for all three risers. However, since the experimental data points were omitted from these maps, the relative differentiation between each flow region cannot be determined and the transition lines as presented are arbitrary.

Das et al. (1999) reported the results of severe slugging experiments in a 10 m high catenary riser in order to determine the effect of compressible gas flow on severe slugging characteristics. Experimental results indicated that increasing gas volume increased (1) the limiting gas velocities for severe slugging, and (2) the cycle time and slug length as a function of gas velocity. This was attributed to reducing the rate of gas pressure accumulation in the pipeline, thereby allowing a longer time liquid slug accumulation.

Nydal et al. (2001) have made a series of experiment for air-water two-phase flow in a flow line-riser system. The riser has an S-shaped form (7m high) and the flow line has a 1-degree downwards inclination towards the riser base (50 mm ID). The flow loop was instrumented with absolute pressure transducers and impedance ring probes for hold up measurement. A stability map showing the approximate boundaries of terrain slugging I (full liquid blocking at the riser base) and terrain slugging II (partial blocking at the riser base) has been generated. Experimental results for increasing gas flow rates at a constant liquid flow rate compared well with predictions. An effect of the S-shaped geometry was demonstrated, in which large slugs appeared to be broken into smaller slugs.

A series of experiments were carried out by Montgomery (2002) with a three-phase mixture of air, oil, and water in an S-shaped riser over the pressure range 2, 4, and 7 bar(a). The test facility consists of a 69 m, 50 mm internal diameter pipeline, which enters the base of the riser at an angle of -2 degree to the horizontal. The riser is 9.9 m tall and runs into a topside separator mounted on a tower. The collected data were used to characterize the unstable flows in terms of pressure cycling, riser liquid inventory, and fluid production characteristics. Experimental results showed that in terms of slug characteristics, transitional severe slugging (including SS2 and SS3) and oscillation flows are as potentially problematic as classical severe slugging, because of the magnitude of peak flow in excess of the average fluid throughput in the riser and the size of the liquid slugs generated.

Yeung and Tchambak (2003) reported the detailed results of severe slugging experiments in an S-shaped riser. The experimental setup and riser geometry are the same as described in Montgomery (2002). However, additional hold up sensors were located at the base of the riser, in the down comer, and the upper limb, and two sets of experiments were carried out under this experimental conditions in order to determine (1) the overall characteristics of the riser for a different air and liquid flows, and (2) the effects of the water cut on severe slugging characteristics. From analysis of the flow behavior, it was found that the riser exhibited all flow regimes as observed for air-water and air-oil two-phase flows. An interesting behavior was that the phase separation could occur in the riser during severe slugging 1 (SS1), where the liquid slugs were segregated into the oil, and water/oil phases. In addition, the liquid slug volume for air/oil two-phase flows was larger than air/water two-phase flows, where the maximum slug volumes occurred at some intermediate water cuts less than 30%. Furthermore, for water cut down to 35%, the maximum pressure difference across the riser was similar to when the water cut is 100%, i.e. the riser is filled with water.

2.3.2. Transient Code Predictions

Comparisons between experiments and code predictions are required to identify the ability of codes to predict the overall flow regime behavior of the riser, including transitional severe slugging and stable flows. This would also identify areas of improvement to the codes, which could lead to better accuracy of predictions. To day, there are only limited studies of comparing severe slugging experiments in flexible risers with transient codes predictions. Kashou (1996) reported the compared results of OLGA code predictions with data generated from the Multiphase Pipeline and Equipment (MPE) joint industrial project. Comparisons have been carried out for two riser configurations (catenary and lazy S) with the outlet pressure boundary specified at atmospheric pressure. Simulated results demonstrated a degree of success in predicting the overall flow regime, cycle times, and slug lengths in the pipeline-riser system; however upon further examination, details of severe slugging characteristics such as peak production rate, the steady production rate, and the pressure cycling characteristics have not been correctly predicted by the code.

Yeung et al. (2003) presented a set of simulation results using experimental data of Cranfield University S-shaped riser, as part of the Transient Multiphase Flows (TMF) projects, in order to (1) determining the effects of the magnitude of boundary variations, and (2) exploring some of the reasons for difference between experimental and simulation results. Simulated results clearly showed that the behavior of the riser is much affected by boundary conditions, where unrealistic conditions can lead to erroneous results. In addition, the liquid hold-up in the down comer was over predicted because of the assumption of a horizontal chordal gas-liquid interface, where in reality the interface is curved.

2.4. REFERENCES

Bendiksen, K.H., "An Experimental Investigation of the Motion of Long Bubbles in Inclined Tubes", *Int. J. Multiphase Flow,* 10, 4, 467-483 (1984).

Bjune, B., Moe, H., and Dalsmo, M., "Upstream Control and Optimization Increases Return on Investment", *World Oil,* 223, 9 (Sept. 2002).

Boe, A., "Severe Slugging Characteristics", course on Selected Topics in *Two-Phase Flow,* NTH, Trondheim, Norway (1981).

Brill, J.P., and Beggs, H.D., "Two-phase Flow in Pipes," 6th Edition, Tulsa University Press, Tulsa, OK (Jan.1991).

Carew, P.S., Thomas, N.H., and Johnson, A.B., "A Physically Based Correlation for the Effects of Power Law Rheology and Inclination on Slug Bubble Rise Velocity", *Int. J. Multiphase Flow,* 21, 6, 1091-1106 (1995).

Das, I., Wordsworth, C., and McNulty, G., "Drilling and Production Technology: Living with Slugs on Floaters", paper presented at the Institute of Marine Engineers Deep and Ultradeep Water Offshore Technology Conference, Newcastle, UK (March 25-26, 1999).

Fabre, J., et al., "Severe Slugging in Pipeline/Riser Systems, *SPE Production Engineering,* 5, 3, 299-305 (Aug., 1990).

Fuchs, P., "The Pressure Limit for Terrain Slugging", Proceeding of the 3rd BHRA International Conference on Multiphase Flow, 65-71, Hague, Netherlands (May 1987).

Furlow, W., "Suppression System Stabilizes Long Pipeline-Riser Liquid Flows", Offshore, Deepwater D&P, 48+166 (Oct., 2000).

Griffith, P., and Wallis, G.B., "Two-Phase Slug Flow", Journal of Heat Transfer, Trans. *ASME,* 82, 307-320 (Aug. 1961).

Harmathy, T.Z., "Velocities of Large Drops and Bubbles in Media of Infinite or Restricted Extent", *AIChE Journal,* 6, 281-288 (1960).

Hatton, S.A., and Howells, H., "Catenary and Hybrid Risers for Deepwater Locations Worldwide", paper presented at Advances in Riser Technologies Conference, Aberdeen, UK (June 1996).

Hatton, S.A., and Willis, N., "Steel Catenary Risers for Deepwater Environments", OTC 8607, Proc. Offshore Technology Conference, Houston, TX (May 1998).

Havre, K., and Dalsmo, M., "Active Feedback Control as the Solution to Severe Slugging", paper presented at SPE Annual Technical Conference and Exhibition, SPE 71540, New Orleans, Louisiana (Oct. 3, 2001).

Havre, K., Stornes, K., and Stray, H., "Taming Slug Flow in Pipelines", ABB Review No. 4, 55-63 (2000).

Hedne, P., and Linga, H., "Suppression of Terrain Slugging with Automatic and Manual Riser Chocking", *Advances in Gas-Liquid Flows,* 453-469 (1990).

Henriot, V., Courbot, A., Heintze, E., and Moyeux, L., "Simulation of Process to Control Severe Slugging: Application to the Dunbar Pipeline", paper presented at SPE Annual Conference & Exhibition, SPE 56461, Houston, TX (1999).

Hill, T.J., "Riser-Base Gas Injection into the S.E. Forties Line", *Proceeding of the 4th BHRA International Conference on Multiphase Flow,* 133-148, Nice, France (June 1989).

Hill, T.H., "Gas Injection at Riser Base Solves Slugging, Flow Problems", *Oil & Gas Journal,* 88, 9, 88-92 (Feb. 26, 1990).

Hollenberg, J.F., DeWolf, S., and Meiring, W.J., "A Method to Suppress Severe Slugging in Flowline/Riser Systems", Proceeding of the 7th BHRG International Conference on Multiphase Flow, 89-103, Cannes, France (June 1995).

Jansen, F.E., "Elimination of Severe Slugging in a Pipeline/Riser System", MSc Thesis, University of Tulsa, Tulsa, OK (1990).

Jansen, F.E., Shoham, O., and Taitel, Y., "The Elimination of Severe Slugging, Experiments and Modeling", *Int. J. Multiphase Flow,* 22, 6, 1055-1072 (1996).

Johal, K.S. et al., "An Alternative Economic Method to Riser Base Gas Lift for Deepwater Subsea Oil/Gas Field Developments", *Proceeding of the Offshore Europe Conference,* SPE 38541, 487-492, Aberdeen, Scotland (9-12 Sept., 1997).

Kashou, S., "Severe Slugging in an S-Shaped or Catenary Riser: OLGA Prediction and Experimental Verification", paper presented at Advances in *Multiphase Technology Conference,* Houston, TX (June 24-25, 1996).

Kovalev, K., Cruickshank, A., and Purvis, J., "Slug Suppression System in Operation", paper presented at the 2003 Offshore Europe Conference, Aberdeen (Sept. 2-5, 2003).

Lunde, O., "Experimental Study and Modeling of Slug Stability in Horizontal Flow", PhD Thesis, The Norwegian Institute of Technology, Trondheim, Norway (1989).

McGuiness, M., and Cooke, D., "Partial Stabilization at St. Joseph", Proceeding of the 3rd International Offshore and Polar Engineering Conference, 235-241, Singapore (June 6-11, 1993).

Molyneux, P., Tait, A., and Kinving, J., "Characterization and Active Control of Slugging in a Vertical Riser", Proceeding of the 2nd North American Conference on Multiphase Technology, 161-170, Banff, Canada (June 21-23, 2000).

Montgomery, J.A., "Severe Slugging and Unstable Flows in an S-Shaped Riser", PhD Thesis, Cranfield University, Bedfordshire, England (Feb. 2002).

Nydal, O.J., Audibert, M., and Johansen, M., "Experiments and Modeling of Gas-Liquid Flow in an S-shaped Riser", Proceeding of the 10th International Conference on Multiphase Flow (Multiphase '01), Cannes, France (June 13-15, 2001).

Pots, B.F.M., et al., "Severe Slug Flow in Offshore Flowline/Riser Systems", paper presented at SPE Middle East Oil Technical Conference & Exhibition, SPE 13723, 347-356, Bahrain (March 11-14, 1985).

Sarica, C., and Shoham, O., "A Simplified Transient Model for Pipeline-Riser Systems", *Chemical Engineering Science*, 46, 9, 2167-2179 (1991).

Sarica, C., and Tengesdal, J. Ø., "A New Technique to Eliminate Severe Slugging in Pipeline/Riser Systems", *Proc. SPE Annual Technical Conference & Exhibition*, SPE 63185, 633-641, Dallas, TX (Oct., 2000).

Schmidt, Z., Brill, J.P., and Beggs, H.D., "Choking Can Eliminate Severe Pipeline Slugging", *Oil & Gas Journal*, 12, 230-238 (Nov. 12, 1979).

Schmidt, Z., Brill, J.P., and Beggs, H.D., "Experimental Study of Severe Slugging in a Two-Phase Flow Pipeline Riser-Pipe System", *SPE Journal*, 20, 407-414 (Oct. 1980).

Schmidt, Z., et al., "Severe Slugging in Offshore Pipeline Riser-Pipe System", *SPE Journal*, 27-38 (Feb.1985).

Schotbot, K., "Methods for the Alleviation of Slug Flow Problems and Their Influence on Field Development Planning", SPE European Petroleum Conference, SPE 15891, London, England (Oct., 18-19, 1988).

Storkaas, E., Alstad, V., and Skogestad, S., "Stabilization of Desired Flow Regimes in Pipelines", paper presented at AIChE Annual Meeting, paper 287d, Reno, Nevada (2001).

Taitel, Y., and Dukler, A.E., "A Model for Predicting Flow Regime Transitions in Horizontal and Near Horizontal Gas-Liquid Flow", *AIChE Journal*, 22, 47-55 (1976).

Taitel, Y., "Stability of Severe Slugging", *Int. J. Multiphase Flow*, 12, 2, 203-217 (1986).

Tchambak. E., "Mitigation of Severe Slugging Using Gas Injection", Interim PhD Review, Report No. 1, Cranfield University, Bedfordshire, England (2003).

Tchambak, E., "Mitigation of Severe Slugging Using Gas Injection", Interim PhD Review, Report No. 2, Cranfield University, Bedfordshire, England (2004).

Tin, V., "Severe Slugging in Flexible Risers", Proceeding of the 5th International Conference on Multiphase Technology, 507-525, Cannes, France (1991).

Tin, V., and Sarshar, S., "An Investigation of Severe Slugging Characteristics in Flexible Risers", Proc. 6th International Conference on Multiphase Production, PP. 205-227, Cannes, France (June 1993).

Vierkandt, S., "Severe Slugging in a Pipeline-Riser System, Experiments and Modeling", MSc Thesis, The University of Tulsa, Tulsa, OK (1988).

Watson, M., Pickering, P., and Hawkes, N., "The Flow Assurance Dilemma: Risk Versus Cost?", Hart's E & P Cover Story (May 2003).

Yeung, H., "Flexible Risers Severe Slugging", Progress Report Number 3, Project 4, Managed Program on Transient Multiphase Flows, Cranfield University, Bedfordshire, England (1996).

Yeung, H., Tchambak, E., and Montgomery, J., "Simulating the Behavior of an S-Shaped Riser", 11th International Multiphase Flow Conference (MULTIPHASE '03), San Remo, Italy (June 11-13, 2003).

Yeung, H., and Tchambak, E., "Three-Phase Flows in an S-shaped Riser – Some Preliminary Results", 11th International Multiphase Flow Conference (MULTIPHASE '03), San Remo, Italy (June 11-13, 2003).

Yocum, B.T., "Offshore Riser Slug Flow Avoidance: Mathematical Model for Design and Optimization", paper presented at SPE London Meeting, SPE 4312, London, UK (1973).

Chapter 3

MODELING OF OFFSHORE PRODUCTION FACILITIES

This chapter discusses how a multiphase flow simulator can be combined with a dynamic process simulation software to reduce uncertainty and minimize losses during operation of the offshore receiving facilities. The HYSYS process simulator is used to build a model of the receiving process facilities and its control system. An OLGA model of the multiphase transport is integrated into this model. This enables the dynamic simulation of both multiphase transport and processing systems.

3.1. INTRODUCTION

In practice, production facilities are often designed in separate stages as isolated sets of equipments. As an example, the multiphase pipelines and separation systems are often designed separate from each other. It is therefore the designer responsibility to use all methods available at the design stage to ensure flow assurance, a controllable process and the integrity of the production facilities. A combined multiphase simulator/dynamic process simulator is a powerful tool for analyzing key questions related to the flow assurance of such systems. By integrating these types of simulators, realistic boundary conditions are obtained for both the multiphase transport pipelines and the receiving process facilities. An integrated transient hydraulic analysis with dynamic process simulation software can also be used to solve and analyze the following problems (Howell, 2001; Sund and Cameron, 2002):

1. Do the production conditions favor the establishment of severe slugging?
2. In the severe slugging region, what is the amplitude and cycle time of slug flow?
3. Does the production control system promote unstable flow regime?
4. Automatic tuning of multiphase pipelines models.
5. Real time modeling of transport and processing systems for providing operator support.

The unique technique used for integrating transient hydraulic flow with process dynamics, shows how the combined approach can begin to address these problems and produces a reliable; real time solution to what is a complex problem that has not been fully tackled in the past.

3.2. ACCEPTANCE OF DYNAMIC SIMULATION

Usually, the operators try not to operate in the severe slugging region. But, the inlet conditions of a production pipeline are linked to the number and the capacity of the producing wells; the availability of wells and also to some undesirable operation such as shut down or restart. The natural trend when dimensioning a production line is to do whatever is possible to avoid critical flooding of the separator, and therefore to over dimension the separator unit. But in offshore production, over dimensioning the installation is very costly and not always possible. So, the design engineers require more accurate transient simulations to correctly design and dimension their production schemes (which are more sensitive to transients occurring when slug flow conditions build-up, and require high performance control systems to maintain the plant within the pre-set operating ranges), and to be able to propose new concepts suitable to every situation they can be faced to (Sagatun, 2004).

Dynamic simulation is the tool of choice to reproduce the expected behavior of still-to-be-built plants, as well as to match the operating conditions of already functioning assets (Feliu et al., 2003). Dynamic simulation provides the analysis of the time history of plant's physical and control variables, which affect system's design and operations management. At the conceptual design stage, dynamic simulation studies are particularly valuable in evaluating process design options, and carrying out controllability studies. During the detailed design phase, dynamic simulation can be used as a tool to assess operability of the process, verify the control philosophy, check and develop start-up/shut-down procedures, and examine case scenarios. Dynamic simulation is also valuable in identifying potential commissioning difficulties, verifying commissioning procedures and establishing control settings and calibration.

3.3. CONCEPT FOR HYSYS-OLGA COUPLING

HYSYS is a tool developed by Hyprotech (1999) for both steady state and dynamic simulation of oil and gas processes. Within HYSYS, steady state simulations can be cast easily into dynamic simulations by specifying additional engineering details, including pressure/flow relationships and equipment dimensions. Control schemes can also be configured within the HYSYS environment from a pre-built suite of function blocks. However, HYSYS does not implement its own tools for transient multiphase flow analysis. It rather uses OLGA2000 simulator, the market-leading simulator for transient multiphase flow of oil, water and gas in pipes, which is licensed separately from Scandpower (the OLGA simulator facilitates transient multiphase simulation features. The governing equations describe mass balances for each of the phases, momentum balances and energy balances. Interactions between the phases are computed based on semi-empirical models. Physical properties of the fluids are pre-tabulated using the PVT-package PVTSIM. OLGA also features a flow regime estimation function that automatically computes the transition between various flow regimes, depending on the pipeline conditions). The HYSYS Dynamics and OLGA simulators run as two separate processes. These processes exchange pressure and flow information as they simulate in parallel. The coupled simulator allows multiphase flow lines and reception facilities to be modeled in a fully integrated way using a single user interface. In fact, this combination enables us (1) to connect multiple HYSYS feed or product streams for modeling of pipeline-riser system with downstream equipments, (2) to assess control systems to help significantly reduce slugging, and (3) to ensure that receiving facilities are properly sized to accommodate any possible slugs or pipeline variations.

3.4. OFFSHORE PRODUCTION PROCESS DESCRIPTION

Figure 3-1 presents a schematic process flow diagram of the offshore production rig. As can be seen, the rig can be divided into three sections: (1) Supply Section, (2) Test Section, and (3) Phase Separation Section. These sections are connected in series in a closed loop operation. The supply section supplies the flow of air, oil, and water to the system. The air is supplied to the system through a compressor. From the compressor outlet, the gas passes through a cooler and then through a large air receiver to reduce the pressure fluctuation from the compressor. The oil and water are stored in two main tanks. The liquids are pumped to the test section from their respective tanks using two pumps. The test section consists of a downward inclined pipeline (flowline) terminated in a riser and a vertical two-phase separator. This separator simulates the first stage of the offshore topside facilities and is used in the mass balance calculation to determine the outlet flow rates of each phase. The separator is in a vertical orientation with air and water exiting from top and bottom of the separator respectively. The air and liquid exiting the separator flow back through separate lines into the three-phase separator. Upon returning from the test section, the fluids pass to the phase-separation section, which comprises a three-phase separator and two coalescers. The three-phase (air/oil/water) mixture enters the main three-phase separator, which is designed to remove the air from the process and carry out the bulk separation of the oil and water. The liquids exiting the separator then pass to the oil and water coalescers respectively for fine fluids separation before returning to their respective tanks. The oil and water can be separated by gravity and by coalescer socks, where the liquid flow through which help smaller drops to coalesce to form bigger drops.

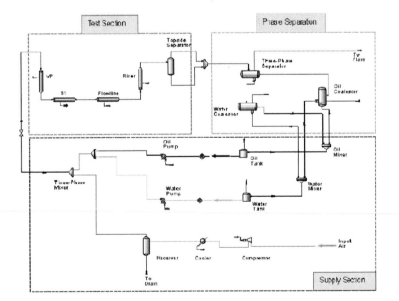

Figure 3.1. Schematic process flow diagram of the offshore production rig.

3.5. DYNAMIC MODEL DEVELOPMENT

3.5.1. Methodology

The HYSYS.Plant (v.3.2) simulator is used to build a model of the downstream (receiving) facilities and its control system. A model to study dynamic behavior of multiphase flow in pipeline-riser system is also developed under the environment of OLGA2000 software, which will then be integrated into the HYSYS model. The coupled simulator allows multiphase flow line and reception facilities to be modeled in a fully integrated way using a single user interface. More information on the above-mentioned simulators can be found in *Appendices A and B*, respectively.

3.5.2. Steady-State Simulation

The starting point for any dynamic simulation study is a sound steady-state simulation. Similarly, a sound appreciation of the steady state behavior of a process is required, as this forms the basis for any control study. Starting with the steady-state simulation model, the necessary equipment information and flowsheet specifications should be setup to permit dynamic simulation analysis. The information, on the initial conditions of the process or time constants[1] of unit operations, needs to be known such that a set of differential equations with respect to time can be solved. Typically, the initial conditions could be obtained from steady-state solutions of the system. The solutions provided temperature, pressure, composition, and flow rates for each stream and each unit operation. To obtain a time constant for a unit operation, one has to know its size. The size of each unit operation was shown in Table 3-1. All separators and coalescers were set to be 50% liquid level.

Before a transition from steady-state to dynamics occurs, the simulation flowsheet should be set up so that a pressure drop exists across the plant. This pressure drop is necessary because the flow in HYSYS Dynamics is determined by the pressure drop throughout the plant. Hence, there is a need to have a valve

---

[1] The time constant defines the speed of the system response to an input, which can be used for control loop analysis.

between every pair of holdup volumes. In this simulation, equal percentage[2] valves are used and sized according to plant data.

Pumps and compressor simulated using HYSYS unit operations based on the operating information specified in the real plant, and the equipment design data, given in Table 3-1.

Table 3.1. The size of unit operations in the process

Equipment	Parameter	Size
Topside Separator	Height	1.400 m
	Diameter	0.480 m
Three-Phase Separator	Length	5.536 m
	Diameter	1.600 m
Oil-Coalescer	Height	2.750 m
	Diameter	0.895 m
Water-Coalescer	Height	2.750 m
	Diameter	0.895 m
Air Receiver	Height	4.000 m
	Diameter	1.516 m

3.5.3. Basic Assumptions

The basic assumptions in the process modeling are presented below:

1. Because the HYSYS simulator currently does not support dynamic unit operations for the pipe segment, it used a valve for the pipe resistance, but this is an approximation and so will not be an accurate representation of the true pipe pressure drop. For this purpose, an OLGA model of the pipeline-riser system needs to be integrated into this model.

2. In real condition, the outflows from the topside separator are connected to the three-phase separator separately, but in the HYSYS simulation since the equipment library does not contain the two-inlet three-phase separator, a mixer has been added in simulation model before three-phase separator to provide single input in the separator.

---

[2] Denotes a percentage change in flow equal to the percentage change in lift or rotation. Valves having this characteristic are more sensitive to plug or vane-position change as the valve nears the wide-open position. Since this is normally the operating position, most control valves are equal percentage.

3. As the HYSYS software does not contain the coalescer in its equipment library, the current study has used two three-phase separators instead of both oil and water coalescers.

4. During simulation, sometimes depending upon the operating conditions, pump cavitation may result from the introduction of gas into the inlet fluid stream, which is an undesirable condition that causes a reduction in pump efficiency and excessive wear or damage to pump components. Hence, for preventing such problem in modeling the process, there is no connection between the two outlet streams from the oil and water coalecsers and two feed streams of pumps at the current process modeling.

3.5.4. Control Strategies

The model also includes a part of the plant control system. With plant-wide control concept, fourteen control loops are required for the current process configuration. These are 2 pressure control loops for topside and three-phase separators, 8 liquid-level control loops for the separators and coalecsers, 3 flow control loops for fluids supply lines. There is also an 'on/off' controller for the compressor as in practice, the compressor doesn't work continuously. It will start to work in order to achieve the desired value of receiver's pressure; however, when the desired pressure of receiver is obtained it will be stopped. The way the 'on/off' controller is configured for the compressor in the model is that it will switch on at 7 barg and then switch off at 7.3 barg, the pressure must then drop to 7 barg before it will switch on again. Considering this type of compressor's operation, the threshold value of 'Full Power' Digital Point sets the compressor switch on pressure and the threshold value of Reset Digital Point sets the compressor switch off pressure. Note, the 'on/off' controller is an appropriate controller if the deviation from the setpoint is within an acceptable range and the cycling does not destabilize the rest of the process.

Control schemes configured within the HYSYS.Plant environment from a pre-built suite of function blocks. Table 3-2 shows the controllers type/action, and the tuning parameters for some important control loops.

Table 3.2. Controller parameters for the system

Control Loop	Controller Type	Controller Action[3]	Controller Gain, K_C	Integral Time, T_i (sec)
FIC-100	PI	Reverse	0.25	6.00
FIC-101	PI	Reverse	0.25	6.00
FIC-102	PI	Reverse	0.25	6.00
PIC-101	PI	Direct	1.08	119.1
PIC-102	PI	Direct	3.28	20.00
LIC-101	PI	Direct	0.77	30.00
LIC-102	PI	Direct	5.00	0.10
LIC-103	PI	Direct	4.15	33.00

Figure 3-2 shows a simplified flowsheet showing the major equipments and basic control strategy of the system.

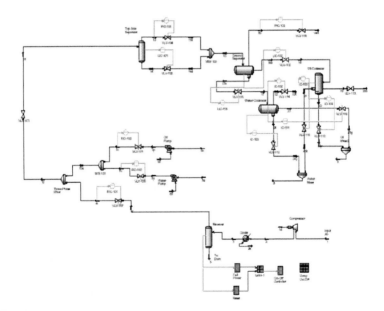

Figure 3.2. Simplified process flowsheet of the offshore production facilities.

---

[3] Controllers can be set up in either direct or reverse modes. If a positive error increases the control output, the controller is said to be direct acting. In the opposite case, when a positive error decreases control output, the controller is said to be reverse acting.

3.5.5. Event Scheduler

To achieve the functional objectives, the Event Scheduler, which is a model for plant events such as pressure excursions, flow changes, and process equipment going down in emergency situations, was also included in the model. A combination of these features, complemented with the custom model of the plant, provides a highly realistic environment for studying real plant behavior and testing the new control strategies and process design options.

3.5.6. Pipeline-Riser Model

As stated previously, a model to study dynamic behavior of multiphase flow in pipeline-riser system is developed under the environment of OLGA2000 software. The model consists of three major parts, fluids PVT description, a pipeline-riser model, and the specifications of the boundary conditions. The PVT behavior is calculated from a model consisting of a list of fluids, their fractions in the mixture, and the temperature and pressure range expected in the system. The fluids consisted of air, oil, and water. The air was simulated using a mixture of nitrogen and oxygen. The oil was also represented as a pseudo component with molecular weight 150.61, liquid density 809 Kg/m^3, normal boiling point 180 °C, critical temperature 200 °C, critical pressure 25.42 bara, and acentric factor 0.424. The Peng and Robinson (1976) equation of state (PR-EOS) is used for the calculation of PVT behavior in all simulations, where PVT table limits of 0.5 to 10 bar (pressure) and 5 to 50 °C are used as input to the PVT calculations. Note, the pressure and temperature range of the simulation must no only cover the range of expected conditions within the pipeline-riser system but also all conditions that the numerical scheme may encounter. Once the values are beyond the range of the PVT table, the thermodynamic properties cannot be associated with the calculation, causing a PVT table error, halting code execution.

The pipeline-riser model consists of a pipe model and a geometry model, which provide generic information to simulate pipeline-riser system. As described in Chapter 3, the pipe is made of 4" nominal bore stainless steel (SS 304) pipe. Table 3-3 gives some pipe properties, which are used in the simulation. The value of pipe roughness was estimated from carrying out single-phase water flow experiments in the pipeline-riser system (see *Appendix C*).

Table 3.3. Pipe properties

Property	Value	Unit
Density	7900	Kg/m3
Heat Capacity	500	J/Kg-°C
Thermal Conductivity	16	W/m-°K
Nominal Diameter	101.6	mm
Thickness	3.05	mm
Roughness	0.045	mm

Note that successful simulations require accurate pipeline geometry, where the flowrate at which the riser base begins to accumulate liquid depends on the pipeline topography (FEESA, 2003). Figure 3-3 shows the pipeline-riser geometry. The geometry model is taken from the tabulated x-y values for the pipeline-riser profile. The x-y coordinates form a series of pipe sections of short length, which are again divided up into individual computational cells. In fact, for having a good model of the flow in the riser region, short section lengths should be used for discretization of the riser and the pipe joining the riser. The structure of the grid is specified such that at least two cells are in each pipe section and that the ratio of cell length from section to section is approximately two. The section length before the riser should be small, large sections will increase the time for liquid blockage in the bend.

As can be seen from Figure 3-3, the entire simulation length also included the 3 m horizontal section of pipe at the top of the riser, entering the topside separator. In fact, recent simulations studies have confirmed that without such a section, simulation results will give unphysical flows exiting the riser, particularly excess liquid fall-back into the riser post blowdown (Nydal et al., 2001; Montgomery, 2002).

To setup pipeline simulations, one must also specify boundary conditions at the inlet and outlet of the pipeline. The inlet and outlet boundary conditions should closely approximate the behavior of the actual upstream and downstream systems if instabilities are to be modeled correctly. Inappropriate boundary conditions can dampen or amplify the fluctuations in the unstable region and will even affect the position of stability boundary. Hence, careful thought must be given to the level of detail required from the study. The boundary conditions are usually either defined as fixed pressure or fixed flowrate, often fixed flowrate at the inlet or fixed pressure at the outlet. This is common practice for simulation studies of multiphase pipelines using the OLGA software (Ek et al., 2002).

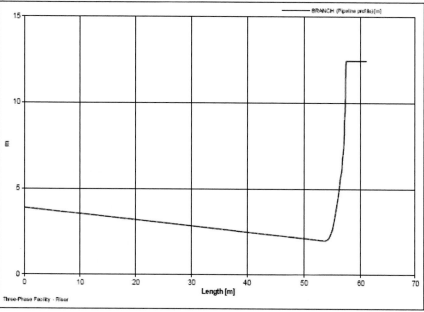

Figure 3.3. Elevation profile for the pipeline-riser system.

Note, transient simulations require initial conditions to begin the simulation. These conditions are then updated as time progresses subject to the boundary conditions and defining equations of the system. In order to assess the stability of the system during normal operation, it is essential to ensure 'fully developed flow'. Consequently, it is important to run the simulation for a significant length of time to ensure that start-up transients, arising from the selection of initial conditions, have decayed leaving the underlying transient behavior. Once fully developed flow has been achieved, the model must be simulated for a sufficient time in order to capture the full periodicity of any surges and to make sure that the system is actually fully developed. It is important however to select the limits of the time-step size carefully. The initial value is set to assure convergence of the simulation in the initial stages of computation. The maximum integration time-step, which should be below the Courant-Friedrich-Levy (CFL) criterion[4] for the

---

[4] In explicit integration methods, the time step is limited by the Courant-Friedrich-Levy criterion based on the speed of sound. Because the speed of sound is about 10^2 to 10^3 times larger than the average phase velocities, explicit integration methods require time steps up to 10^3 times smaller than implicit methods.

flow conditions encountered in the simulation, is set to prevent the PVT table errors, while the minimum integration time step is set to ensure practical simulation times (Bendiksen et al., 1991; Montgomery, 2002).

3.6. REFERENCES

Bendiksen, K.H., Malnes, D., Moe, R., and Nuland, S., "The Dynamic Two-Fluid Model OLGA: Theory and Application", *SPE Production Engineering,* 171-180 (May 1991).

Boe, A., "Severe Slugging Characteristics", course on Selected Topics in Two-Phase Flow, NTH, Trondheim, Norway (1981).

Ek et al., "The Field Simulator-Combined Multiphase Flow and Process Simulation", paper presented at the 3rd North American Conference on Multiphase Technology, Banff, Alberta, Canada (June 6-7, 2002).

Fabre, J., et al., "Severe Slugging in Pipeline/Riser Systems, *SPE Production Engineering,* 5, 3, 299-305 (Aug., 1990).

FEESA, "Life of Field Stability of a North Sea Oil Development", Feesa Ltd Case Study, Surrey, UK (2003).

Feliu, J.A., Grau, I., Alos, M.A., and Macias-Hernandez, J.J., "Match Your Process Constraints Using Dynamic Simulation", *Chemical Engineering Progress,* PP. 42-48 (Dec. 2003).

Howell, A., "Is Your Platform Control System Stressed? – Coupling Transient Pipeline and Process Simulation", paper presented at GPA Europe Meeting, Norwich (May 2001).

Hyprotech, Dynamic Modeling, *Hysys.Plant Manual,* PP. 1-5 (1999).

Montgomery, J.A., "Severe Slugging and Unstable Flows in an S-Shaped Riser", PhD Thesis, Cranfield University, Bedfordshire, England (Feb. 2002).

Nydal, O.J., Audibert, M., and Johansen, M., "Experiments and Modeling of Gas-Liquid Flow in an S-shaped Riser", Proceeding of the 10th International Conference on Multiphase Flow (Multiphase '01), Cannes, France (June 13-15, 2001).

Peng, D.Y., and Robinson, D.B., "A New Two-constant Equation of State", *Ind. Eng. Chem. Fundam.,* 15, 58-64 (1976).

Sagatun, S.I., "Riser Slugging: A Mathematical Model and the Practical Consequences", *SPE Production & Facilities Journal,* 19, 3, 168-175 (2004).

Sarica, C., and Shoham, O., "A Simplified Transient Model for Pipeline-Riser Systems", *Chemical Engineering Science,* 46, 9, 2167-2179 (1991).

Schmidt, Z., et al., "Severe Slugging in Offshore Pipeline Riser-Pipe System", *SPE Journal,* 27-38 (Feb.1985).

Sund, E.B., and Cameron, D., "Design Engineering, Operator Training and Monitoring for Tie-ins and Multiphase Gas Pipelines Using Combined Dynamic Simulation and Multiphase Flow Modeling", paper presented at GPA Europe Meeting, Bergen, Norway (May 2002).

Taitel, Y., "Stability of Severe Slugging", *Int. J. Multiphase Flow,* 12, 2, 203-217 (1986).

Chapter 4

SIMULATION RESULTS AND CONCLUSIONS

This chapter presents the results of a dynamic simulation study of the offshore production rig during different incidents. The study is performed using two separate dynamic simulators HYSYS.Plant and the multiphase flow simulation tool OLGA. Comparisons are made between the experimental results and transient codes. The motive for this work is to identify areas that pose substantial difficultly to the codes and to identify potential methodologies for the simulation of severe slugging in a catenary-shaped riser. The description below is broken into three sections, the first section covers the experimental test cases, the second section covers the simulation results, and finally the third details some numerical experiments undertaken, which examines the sensitivity of the OLGA model's results to computation parameters.

4.1. DESCRIPTION OF THE EXPERIMENTAL CAMPAIGN

The objective of the experimental campaign was to identify performance and capability of the models and to identify, if any, areas of improvement to the codes which could lead to better accuracy of predictions.

The campaign covered two sets of experiments for different inlet air-water flowrates. These were conducted at constant topside separator pressure (1 barg). For the first experiment, the inlet air flowrate was fixed at 10 Sm3/h, and the inlet water flowrate was changed as shown in Figure 4-1. Continuing from the final step of the first experiment, Test 2 was conducted with the fixed water flowrate at 0.5 lit/s and variable air flowrates as given in Figure 4-2. Note, the fluctuations in the measured inlet air and water flowrates for both Tests 1 and 2 are most likely caused by the fluctuations in the topside separator pressure.

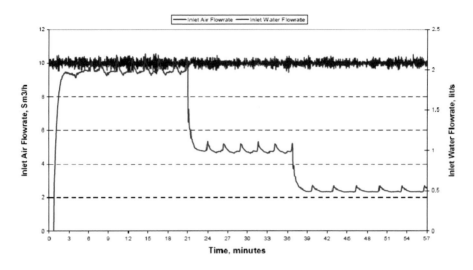

Figure 4.1. Inlet water and air flowrate profiles investigated in the first test of the experimental campaign.

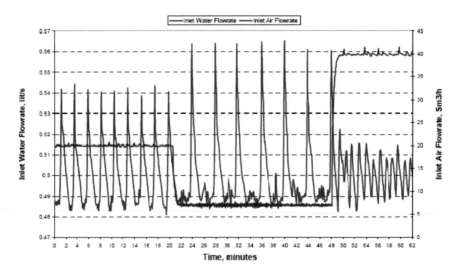

Figure 4.2. Inlet water and air flowrate profiles investigated in the second test of the experimental campaign.

4.2. SIMULATION RESULTS

In the following section, the predicting results of the OLGA model of pipeline-riser system and the HYSYS model of downstream receiving facilities, compared to the experimental data, are shown for the relevant campaign.

4.2.1. Pipeline-Riser System

Figures 4-3 and 4-4 show the OLGA simulation results and experimental data for the riser-base pressure in both Tests 1 and 2 of the experimental campaign. The OLGA simulations have been carried out with a constant pressure boundary equal to the average topside separator pressure (1 barg). For these cases, the pressure difference between maximum and minimum riser-base pressure for the model compared to the experimental data does not agree very well. The OLGA model also shows a severe slugging frequency that is slightly higher than that of the experimental data. However, comparisons between experimental data and computational results for varying water flowrates at a constant air flowrate (Figure 4-3) show good correspondence at low water flowrates (about 1.0 lit/s). There is still a slight time shift between experimental and simulation results at lower water flowrates (0.5 lit/s). At higher water flowrates (about 2.0 lit/s), full blocking occurs, trapping the gas in the pipeline and causing build up of liquid in the riser, and an increasing slugging period. Also for Test 2 (as shown in Figure 4-4), at high air flowrates (20 Sm^3/h), the slugging periods become difficult to estimate as the flow changes to severe slugging II, where there is no slug production period (indicating that no liquid is backed up the pipeline when the severe slugging is fully formed) and the cycle time of severe slugging is also lower than that of severe slugging I, increasing the slugging frequency. The continuous air penetration at the riser-base indicates that there is a lack of gas compression in the pipeline that not allowing the pure liquid formation in the riser. This then may suggest the OLGA model is not predicting the pipeline behavior of the system accurately in terms of gas accumulation. Finally, at higher air flowrates (about 40 Sm^3/h) the flow changes to slug flow, where the maximum pressure difference during slug flow is less than that experienced during severe slugging I, indicating that the riser is not completely filled with liquid. During slug flow, the frequency of the pressure difference fluctuation is also dependent on the frequency at which slugs enter/leave the riser and the magnitude of the fluctuation is dependent on the size of slugs. As can been seen, the OLGA model is also unable to predict slug flow in the pipeline-riser system, where it predicted bubbly flow in the riser rather than slug flow (characterized by stable riser-base pressure profile). The prediction of a bubble flow regime indicates that the flow regime transition mechanism in the OLGA code is not functioning in the region close to the riser-base. This is again attributable to inaccurate prediction of the pipeline gas accumulation behavior, corresponding to over-prediction of the inlet gas superficial velocities as shown in Figures 4-5 and 4-6.

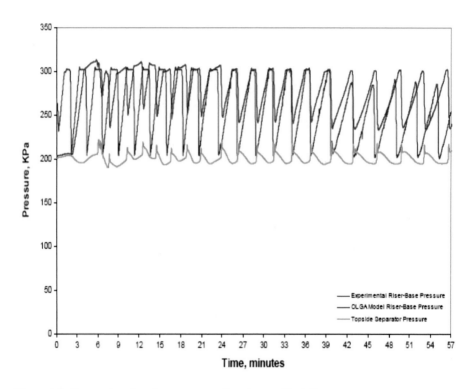

Figure 4.3. Comparing riser-base pressure time traces (Test 1).

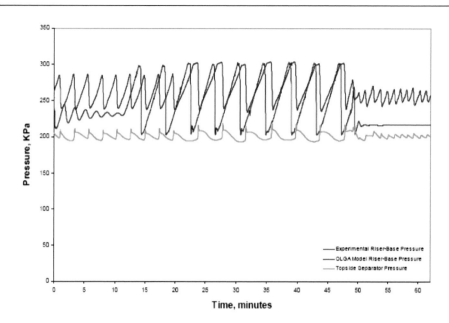

Figure 4.4. Comparing riser-base pressure time traces (Test 2).

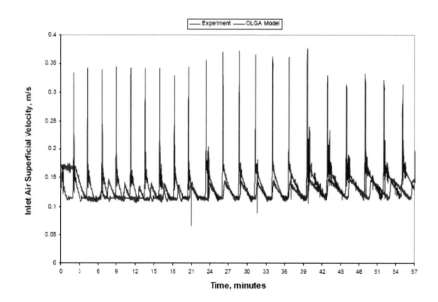

Figure 4.5. Comparing inlet air superficial velocity changes (Test 1).

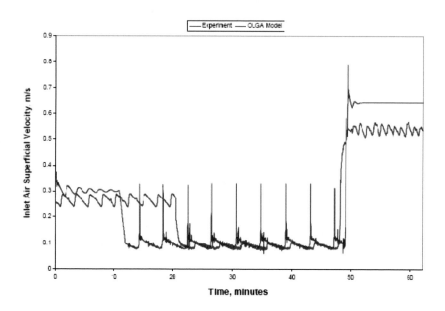

Figure 4.6. Comparing inlet air superficial velocity changes (Test 2).

4.2.1.1. Effects of Magnitude of Pressure Boundary Variation

As stated previously, the outlet boundary condition in both tests was simulated as a constant pressure boundary. However, in reality, this boundary condition was varied because of the imperfect controls and system interactions, causing fluctuations in the measured inlet air and water flowrates, especially during severe slugging flow. Unrealistic boundary conditions have been therefore led to erroneous results. A numerical experiment has been carried out to examine the effect of outlet boundary condition variations on the simulation results. Figure 4-7 and 4-8 show typical riser-base pressure profiles for a constant topside separator pressure, and ±3 % of the averaged value. The inlet air and water flowrates are the same with previous simulation. As can be seen from Figures 4-7 and 4-8, the riser-base pressure behavior is much affected by the pressure boundary condition. Hence, the boundary conditions have to be accurately represented in any simulation results. For this purpose, a tight integration of the pipeline-riser model with the model of the receiving plant may be very important for achieving correct simulation results, especially for studying slugging dynamics.

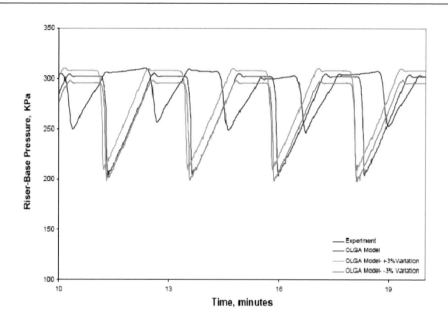

Figure 4.7. Effects of magnitude of pressure boundary variations on the riser-base pressure (Test 1).

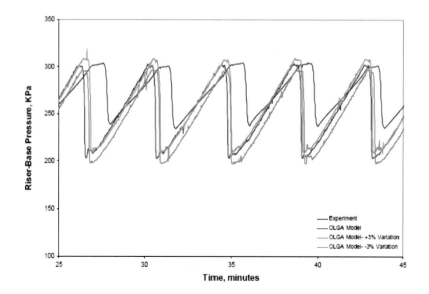

Figure 4.8. Effects of magnitude of pressure boundary variations on the riser-base pressure (Test 2).

4.2.1.2. Flow Regimes Characteristics

The OLGA model is moderately successful in predicting the flow regime in the pipeline-riser system during severe slugging I. However, in terms of the severe slugging pressure cycling characteristics, the model globally under-predicts the cycle time. In each case, the predicted liquid buildup and slug production times are different to those observed. The liquid slug buildup period is longer and the slug production period is shorter than experimental values. This difference is reproduced in Figures 4-9 and 4-10, for each test, showing a comparison between the experimental results and the model prediction. For Test 1, the liquid buildup period from the model is 1.22 min compared an experimental value of 0.58 min. Furthermore, the model-predicted slug production period is 0.88 min compared to 1.18 min for the experiments. In both Tests 1 and 2, the main source of error between the simulation results and the experimental data in terms of the cycle time is the liquid buildup stage of the severe slugging, which may be due to inaccurate prediction of the pipeline liquid holdup and gas compression behavior during severe slugging in the model prediction.

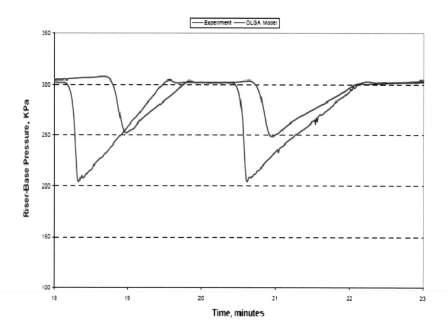

Figure 4.9. Comparing experimental and simulated riser-base pressure profiles during severe slugging I (Test 1).

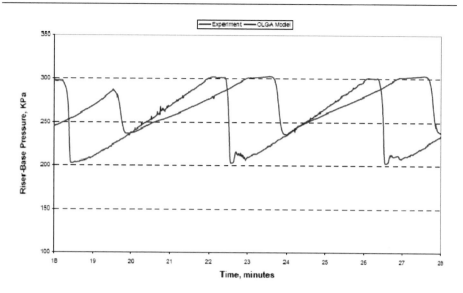

Figure 4.10. Comparing experimental and simulated riser-base pressure profiles during severe slugging I (Test 2).

In terms of fluid production, the trend associated with the pressure cycling is reflected in the liquid production characteristics. Typical examples of liquid mass production during the experiments (both Tests 1 and 2), obtained from the topside separator mass balance, are shown in Figures 4-11 to 4-13. As can be seen from Figure 4-11, the severe slugging I is characterized by three main periods, the period of no production, the period of constant liquid production, and the transient production period. In the first one, the liquid accumulates in the low point due to its low velocity and to the liquid that falls down, and forms a slug until the pressure becomes sufficient to lift the liquid column. In the second one, the liquid slug starts to go upward along the riser. The gas begins to flow in the riser and so accelerates the liquid. Finally, the gas arrives at the top of the riser and the pressure rapidly decreases causing liquid flow down.

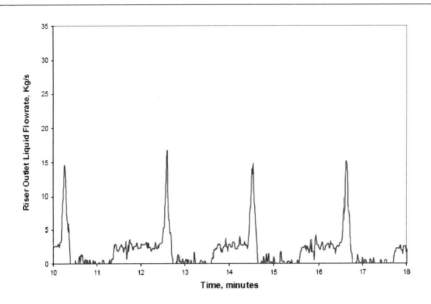

Figure 4.11. Liquid production profile during severe slugging I (Test 1).

Figure 4-12 shows an example of the liquid production profile during severe slugging II. As stated before, there is no constant production period during severe slugging II, therefore, all fluid production in this flow regime is in the form of a single liquid transient production, which occurs during the gas blowdown period of severe slugging.

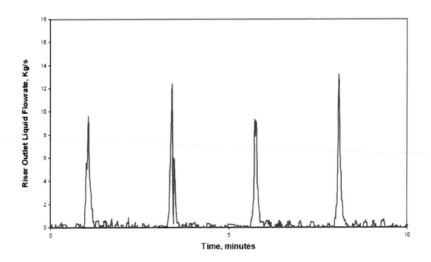

Figure 4.12. Liquid production profile during severe slugging II (Test 2).

Finally, the liquid production profile for slug flow in shown in Figure 4-13. As can be seen, during slug flow, there is continuous delivery of liquids as slug flow is made up of a bubbly liquid and a long bubble/film region. However, there are surges in the liquid production, corresponding to the arrival of gas bubbles.

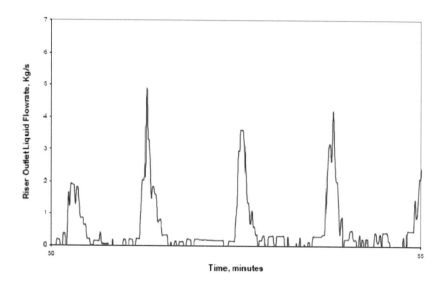

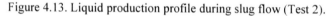

Figure 4.13. Liquid production profile during slug flow (Test 2).

Examining the predicted liquid production profiles during severe slugging I for both Tests 1 and 2 (Figures 4-14 and 4-15), there is a marked difference between the predicted and experimental liquid production profiles. Shorter cycle times in simulation results lead to higher instantaneous outlet liquid flowrate and smaller slug size than observed during experiments. It is also found that the simulations over-predict the peak liquid production rate during severe slugging I, which may be due to incorrect prediction of the rate of bubble propagation through the riser and the gas/liquid interface position in the pipeline. As can be seen from Figure 4-15, the OLGA model is also unable to predict the severe slugging II and slug flows, due to over-prediction of the bubble penetration at the riser-base.

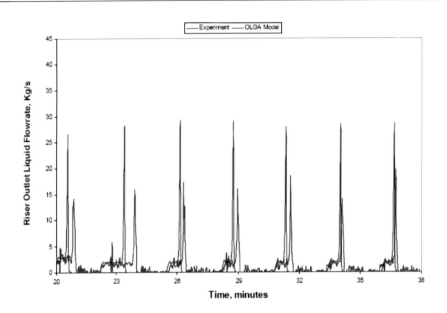

Figure 4.14. Comparing experimental and simulated liquid production time traces (Test 1).

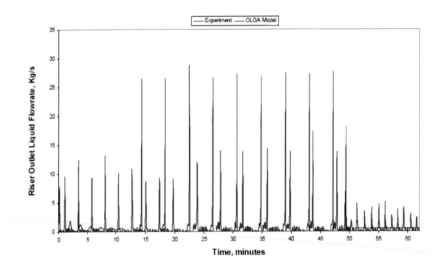
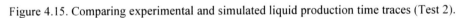

Figure 4.15. Comparing experimental and simulated liquid production time traces (Test 2).

In terms of gas production, typical results for both Tests 1 and 2 are shown in Figures 4-16 to 4-18. As can be seen, during severe slugging I, the gas production cycle is made of two parts, the period of no gas production (which occurs during the liquid buildup, slug production, and bubble penetration stages of the severe slugging cycle) and transient gas production. However, for severe slugging II, the gas production is in the form of a transient gas production followed by a period of constant gas production. Finally, during slug flow, there is continuous delivery of gas as slug flow is made up of a bubbly liquid and a long bubble/film region. However, there are surges in gas production, corresponding to the arrival of slug body.

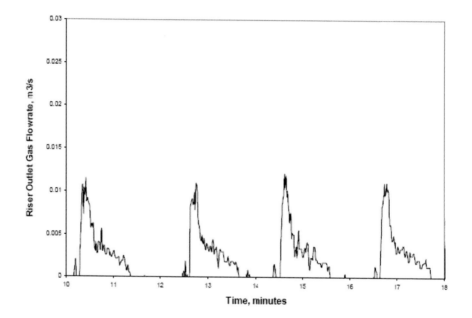

Figure 4.16. Gas production time trace during severe slugging I (Test 1).

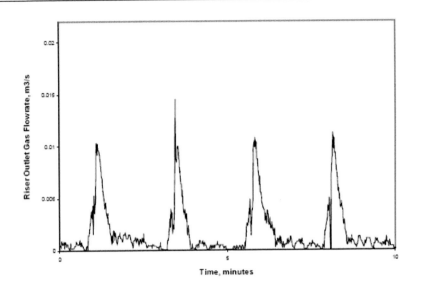

Figure 4.17. Gas production time trace during severe slugging II (Test 2).

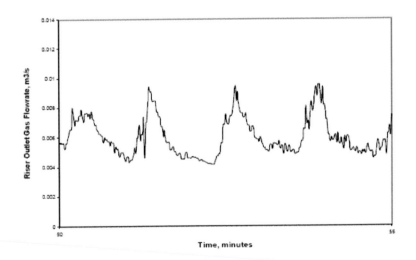

Figure 4.18. Gas production time trace during slug flow (Test 2).

Similarly to liquid production profile, the model predictions of peak flow in gas production, for both cases, are vastly different from the experimental results (Figures 4-19 and 4-20). This is again attributable to inaccurate prediction of the riser-base bubble penetration, pipeline gas accumulation and the gas blowdown.

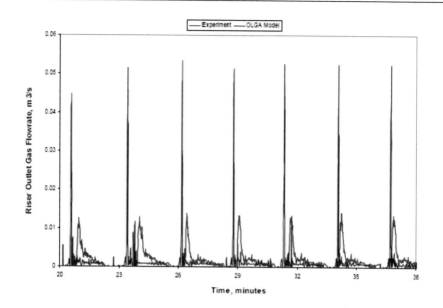

Figure 4.19. Comparing experimental and simulated gas production time traces (Test 1).

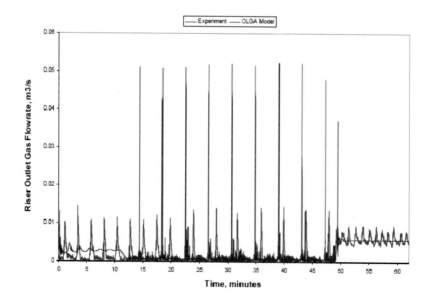

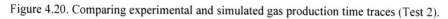

Figure 4.20. Comparing experimental and simulated gas production time traces (Test 2).

4.2.1.3. Effects of the Pipe Grid Density Variation

In order to examine the sensitivity of the OLGA simulation results to computational parameters, a numerical experiment using a variation of the grid density was also carried out[1]. Attempting to improve the simulation results using an increase in grid density was of limited success. As an example, for Test 1, the original simulation predicted a severe slugging I cycle time of 171.24 s, 95% of the experimental value. Increasing the grid density by a factor of 2X of the original density (X), the result is a cycle time of 170.22 s, 94% of the experimental value, a further reduction compared to the experimental value (Figure 4-21).

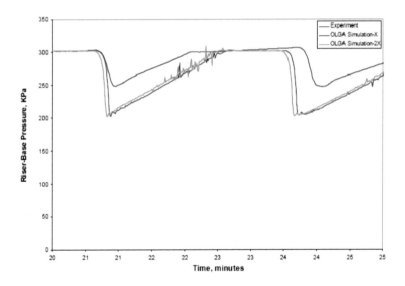

Figure 4.21. Effect of grid density changes on the severe slugging I cycle time (Test 1).

The effect of the shorter cycle on the predicted fluid production characteristics is a marginal reduction in the predicted duration of the constant production period from 54.06 s (45% of the experimental value) to 49.44 s (41% of the experimental value). These changes have a proportional effect of the peak production rates, as shown in Figures 4-22 and 4-23. Comparing the liquid production profiles, it is clear that for most part, the effect of the grid modification is marginal in the order of percent and hence negligible. However, Figure 4-23 shows sensitivity in the gas production rate value to changes in the grid density. The changes in the gas peak production are again attributed to the bubble

---

[1] Since this simulation is rather simple, then computation time would not be that significant.

propagation through the curved riser, which is not predicted correctly. Hence, it may be asserted that the gas accumulation and the gas-liquid interface behavior in the pipeline is highly dependant on the grid density for all simulations.

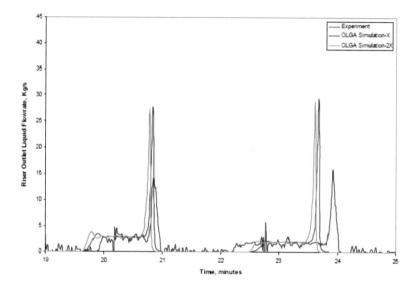

Figure 4.22. Effect of grid density changes on the liquid production profiles during severe slugging I (Test 1).

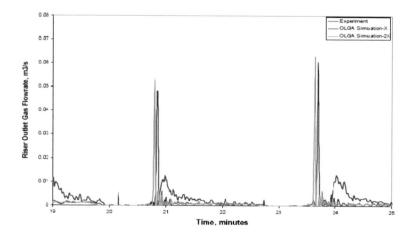

Figure 4.23. Effect of grid density changes on the gas production profiles during severe slugging I (Test 1).

4.2.1.4. Severe Slugging Region Prediction

Figures 4-24 and 4-25 show the Boe (1981) prediction model, which is commonly used to predict severe slugging, compared with the experimental data acquired. The Boe criterion line which has been plotted here is a combination of this criterion and the critical value of liquid velocity, where above this value, the flow pattern in the pipe is dispersed or nearly liquid. Therefore, the severe slugging phenomenon cannot exist (Schmidt et al., 1985). The pipeline liquid holdup in this calculation has been also estimated based upon the model of Taitel (1986) and the computation method developed by Yeung (1996). More information on these methods can be found in *Appendix D*.

In these figures, the Boe (1981) severe slugging region and the operation points are shown. Results show that the Boe stability criterion is unable to predict the severe slugging region in the pipeline-riser system. However, additional validation data is required to confirm this, particularly in respect to different system operating pressures. No comparisons have been given for the above operating conditions between experiments and the model by Fabre et al. (1987) possibly due to the inability of their model to handle the discontinuities of two-phase flow that occur during the slug formation at the bottom of the riser (Sarica and Shoham, 1991).

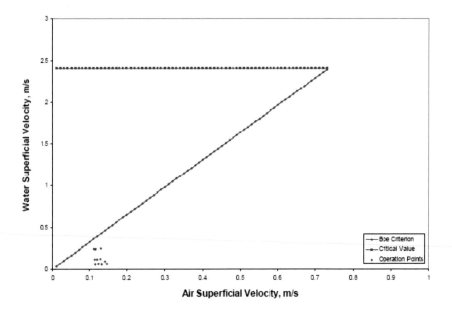

Figure 4.24. Comparing Boe (1981) stability line with experimental data (Test 1).

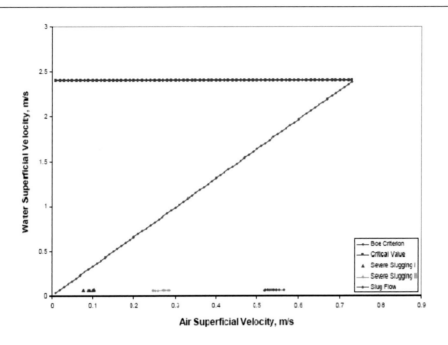

Figure 4.25. Comparing Boe (1981) stability line with experimental data (Test 2).

4.2.2. Downstream Receiving Facilities

Setting up all specifications necessary for running in the dynamic mode, the process plant was simulated dynamically as a control loop system using the same initial conditions as described previously. Topside and three-phase separators were set to be 0.7 m and 65% liquid level, respectively. The coalescers were also set to be 50% liquid level. The pressure in both topside and three-phase separators was set to be 1 barg. Figures 4-26 to 4-29 show the pressure responses at the topside and three-phase separators for both Tests 1 and 2. As can be seen, the separators pressures were disrupted suddenly and returned to their setpoint rapidly. Figures 4-30 to 4-33 also show that liquid level profiles in both topside and three-phase separators are changed for a short period of time and returned to their specified conditions. In fact, in both cases, the control system can move the process to the specified conditions.

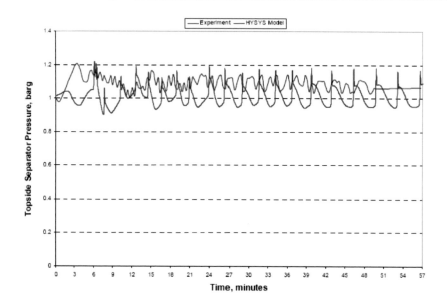

Figure 4.26. Comparing pressure responses at the topside separator (Test 1).

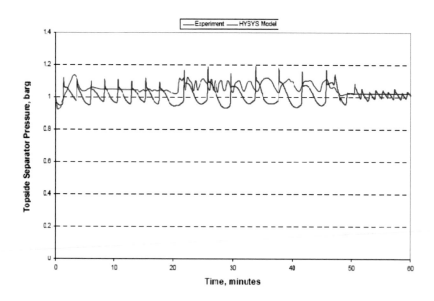

Figure 4.27. Comparing pressure responses at the topside separator (Test 2).

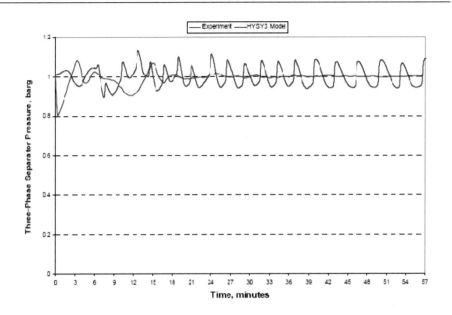

Figure 4.28. Comparing pressure responses at the three-phase separator (Test 1).

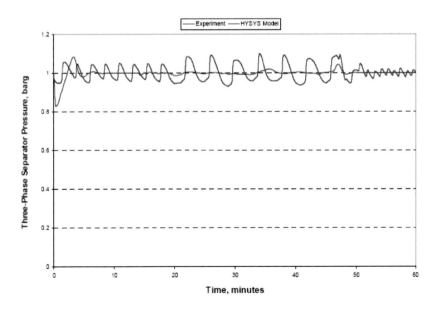

Figure 4.29. Comparing pressure responses at the three-phase separator (Test 2).

70   Saeid Mokhatab

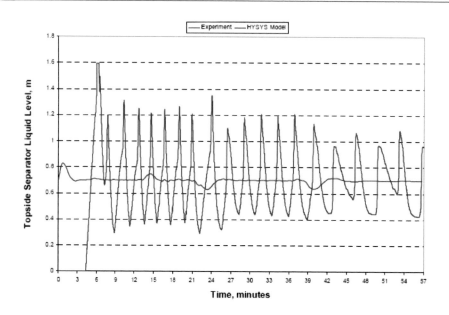

Figure 4.30. Comparing liquid level profiles of topside separator (Test 1).

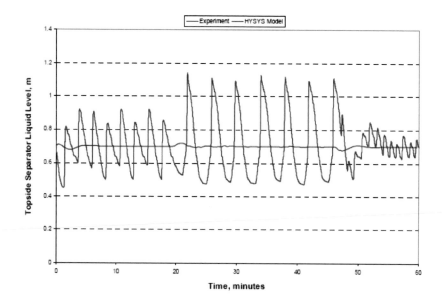

Figure 4.31. Comparing liquid level profiles of topside separator (Test 2).

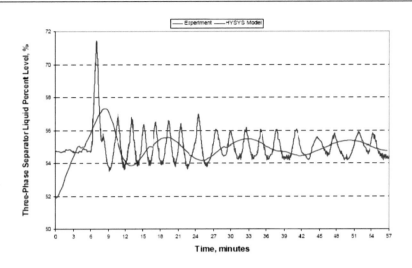

Figure 4.32. Comparing liquid level profiles of three-phase separator (Test 1).

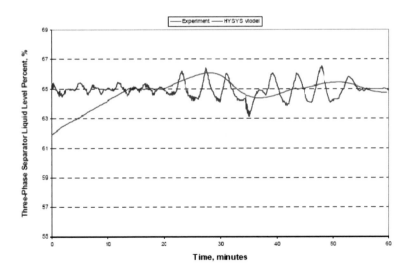

Figure 4.33. Comparing liquid level profiles of three-phase separator (Test 2).

Figures 4-34 to 4-37 show the liquid and gas production profiles from the topside separator for both Tests 1 and 2, and illustrate how the dynamically controlled valves at the gas and liquid outlets upstream of the topside separator suppress surges of outlet flows. In fact, implementation of the outlet control valves results in a stabilized liquid and gas production rates, approximating the ideal production systems.

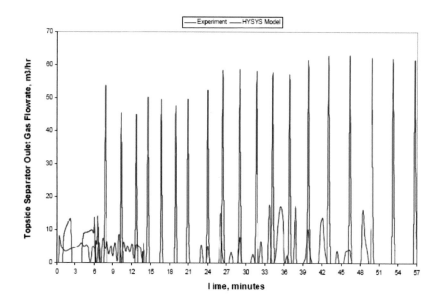

Figure 4.34. Comparing topside separator's gas production profiles (Test 1).

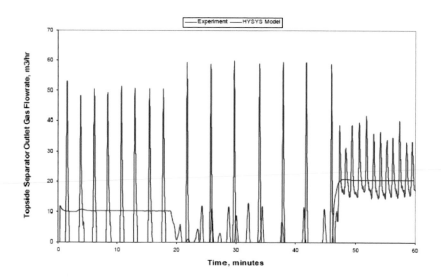

Figure 4.35. Comparing topside separator's gas production profiles (Test 2).

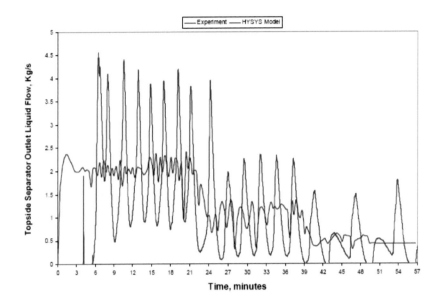

Figure 4.36. Comparing topside separator's liquid production profiles (Test 1).

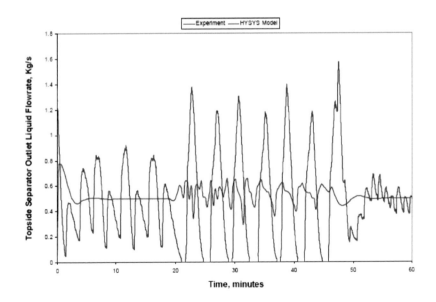

Figure 4.37. Comparing topside separator's liquid production profiles (Test 2).

4.3. CONCLUSIONS

Drawing all simulation results together, the following general observations are made:

1. Considering the external factors such as calibration, compressor and pump effects in the experiments, the OLGA results compare rather well with the experiments.
2. An OLGA model of pipeline-riser system is able to predict the occurrence of the classical severe slugging (SSI) in the test series. However, the detailed results show that the model does not give acceptable prediction of the characteristics of the observed classical severe slugging.
3. The OLGA model has difficulty in predicting the severe slugging II with respect to the bubble penetration at the riser base and the blowdown process. The model is also unable to predict the slug flow in the pipeline-riser system, due to the presence of bubble flow at riser-base. This is attributable to inaccurate prediction of the pipeline gas compression behavior, corresponding to over-prediction of inlet gas superficial velocities.
4. The HYSYS model is able to investigate the dynamic behavior of downstream reception facilities of the Three-Phase Production Rig. However, there remain some errors in predicting system's dynamic behavior due to the HYSYS dynamic initialization[2]. Most of these errors occur on holdup terms. This problem sometime happens during switching a converged HYSYS model from Steady State mode to Dynamic mode, where HYSYS simulator takes some initial values internally, instead of using the values in Steady State mode. However, this problem does not affect the overall dynamic study and the calculation backs to normal once the transition from the initialization is done.

4.4. SUGGESTIONS FOR FUTURE INVESTIGATIONS

1. As stated previously, the prediction of pipeline simulation respect to slug dynamics will be strongly affected by the specification of boundary

---

[2] Similar problem has been reported before and the HYSYS development team is working on the improvement (Wang, 2005).

conditions. Hence, a tight integration of the OLGA model of pipeline-riser system with the HYSYS model of the receiving plant may be very important for achieving correct simulation results, especially for studying slugging dynamics.

2. In this study, severe slugging in a catenary riser has only been characterized in terms of pressure fluctuations at the riser-base and fluids production for a minimum amount of experimental data. Hence, additional experiments should also be carried out in order to exploring the operating characteristics of the pipeline-riser system. These data should therefore cover a range of back pressures and inlet air and liquid velocities in order to establish a new stability map and examine the effects of pressure on the flow regime stability and occurrence of severe slugging.

3. Since the use of existing criteria for severe slugging is usually open to debate, a new stability criterion for the occurrence of severe slugging in a catenary riser should be therefore developed and the obtained results must be compared with the experiments and the results of existing criteria for severe slugging.

4. It is important that the PVT description of the fluids used in the simulation is as representative of the actual fluids as possible, especially for field matching purposes, where the physical properties of the fluids does indeed influence the simulation results. Hence, there is a certain need to have the actual physical property data for the oil (used in the experiments) from the vendor, to extend the sensitivity study on the effect of oil density on the slugging phenomenon.

4.5. REFERENCES

Boe, A., "Severe Slugging Characteristics", course on Selected Topics in Two-Phase Flow, NTH, Trondheim, Norway (1981).

Fabre, J., et al., "Severe Slugging in Pipeline/Riser Systems, *SPE Production Engineering,* 5, 3, 299-305 (Aug., 1990).

Sarica, C., and Shoham, O., "A Simplified Transient Model for Pipeline-Riser Systems", *Chemical Engineering Science,* 46, 9, 2167-2179 (1991).

Schmidt, Z., et al., "Severe Slugging in Offshore Pipeline Riser-Pipe System", *SPE Journal,* 27-38 (Feb.1985).

Taitel, Y., "Stability of Severe Slugging", *Int. J. Multiphase Flow,* 12, 2, 203-217 (1986).

Yeung, H., "Flexible Risers Severe Slugging", Progress Report Number 3, Project 4, Managed Program on Transient Multiphase Flows, Cranfield University, Bedfordshire, England (1996).

Wang, T.J., Personal Communications, Technical Support Specialist, *Aspen.Tech.* Inc., Calgary, Alberta, Canada (2005).

APPENDIX A

HYSYS.PLANT SIMULATOR

The dynamic behavior of the process system is best suited under the environment of HYSYS.Plant, a commercial dynamic simulation tool developed by Hyprotech (2001), which provides accurate results based on rigorous equilibrium, unit operations, and controller models.

Conservation Relationships

The conservation relationships are the basis of mathematical modeling in HYSYS.Plant. HYSYS.Plant uses a different set of conservation equations which account for changes occurring over time. Dynamic behavior arises from the fact that many pieces of plant equipment have some sort of material inventory or holdup. The lagged response that is observed in any unit operation is the result of the accumulation of material, energy, or composition in the holdup. The heat loss experienced by any pieces of plant equipment is also considered by the holdup model in HYSYS. The heat loss model influences the holdup by contributing an extra term to the energy balance equation. Heat is lost or gained from the holdup fluid through the wall and insulation to the surroundings. There are, however, calculations (vessel level and vessel pressure calculations) which are not handled by the holdup model itself, but can impact the holdup calculations. These calculations require more information and are solved in conjunction with the holdup model.

Pressure-Flow Solver

Almost every unit operation in the process flowsheet can be considered a holdup or carrier of material (pressure) and energy. A network of pressure nodes can therefore be conceived across the entire simulation case. The pressure-flow (PF) solver considers the integration of pressure-flow balances in the flowsheet. There are two basic equations, which define most of the pressure-flow network and these equation only contain pressure and flow as variables:

1. Volume balance equations, which define the material balance at pressure holdups.
2. Resistance equations, which define flow between pressure holdups.

The pressure-flow balances both require information from and provide information to the holdup model. While the holdup model calculates the accumulation of material, energy, and composition in the holdup, the pressure-flow solver equations determine the pressure of the holdup and flowrates around it. The holdup model brings the actual feed and product stream properties to holdup conditions for the volume balance equations using a rigorous or approximate flash. The pressure-flow solver returns information essential to the holdup model calculations, the pressure of the holdup or the flowrates of streams around the holdup. The volume balance equations, resistance equations, and pressure-flow relation equations make up a large number of equations in the pressure-flow matrix. To satisfy the degrees of freedom of the pressure-flow matrix, a certain number of pressure-flow specifications should be therefore set.

Volume Balance Equation

For equipment with holdup, an underlying principle is that the physical volume of the vessel, and thus, the volume of material in the vessel at any time remain constant. Therefore, during calculations in dynamics, the change in volume of the material inside the vessel is zero. As such, a vessel pressure node equation is essentially a volumetric flow balance and can be expressed as follows:

$$\frac{dV}{dt} = \frac{d}{dt} f(\dot{m}, h, P, T) = 0 \qquad (1)$$

Where, V is the volume of the vessel; $\dot{m}$, mass flowrate; h, holdup; P, and T are the vessel pressure and temperature, respectively. In the volume balance equation, pressure and flow are the only two variables to be solved in the matrix. All other values in the equation are updated after the matrix solves. Each vessel holdup contributes at least one volume balance equation to the pressure-flow matrix. When sufficient pressure-flow specifications are provided, any unknowns can be solved whether it is a vessel pressure or one of its flowrates.

Resistance Equations

Flows existing from a holdup are calculated from a volume balance equation or calculated from a resistance equation. In general, the resistance equation relates the pressures of two nodes and flow that exist between the nodes. The resistance equations are modeled after turbulent flow equations and have the following form, which is a simplified form of the basic valve operation equation:

$$\dot{m} = k\sqrt{\Delta P} \qquad\qquad (2)$$

Where, $\dot{m}$ is the mass flowrate, k is the conductance (which is a constant representing the reciprocal of resistance to flow), and ΔP is the frictional pressure loss, which is the pressure drop across the unit operation without static head contributions.

HYSYS Model of the Offshore Production Rig

A more detailed description on the individual unit operations present in the HYSYS Dynamic Model of the Three-Phase Facility and the resistance equations associated with them is discussed in the following section.

Air Flow Supply Line

Air flow into compressor is governed by the speed of the compressor as (HYSYS.Plant Manual, 2001):

$$F = \left[\left(1 - L\right) - C \left[\frac{Z_s}{Z_d} \left(\frac{P_d}{P_s} \right)^{\frac{1}{K}} - 1 \right] \right] \left[\frac{\frac{N}{60} \times \rho_{air}}{29} \right] \qquad (3)$$

where:

F: air molar flow, Kmole/hr

L: effects of variable such as internal leakage, gas friction, pressure drop through valves, and inlet gas preheating. The value of L varies from 0.04 to 0.15 in general.

C: clearance volume

Z_s: suction compressibility factor

Z_d: discharge compressibility factor

P_s: suction pressure, psia

P_d: discharge pressure, psia

K: heat capacity ratio, Cp/C_V

N: speed, rpm

ρ_{air} : air density, lb_m/ft^3

The performance of the compressor is also evaluated based on the volumetric efficiency and brake horsepower. Volumetric efficiency, VE, is defined as the actual pumping capacity of a cylinder compared to the piston displacement volume. It is given by (HYSYS.Plant Manual, 2001):

$$VE = \left[\left(1 - L\right) - C \left[\frac{Z_s}{Z_d} \left(\frac{P_d}{P_s} \right)^{\frac{1}{K}} - 1 \right] \right] \qquad (4)$$

To account for losses at the suction and discharge valve, an arbitrary value about 4% VE loss is acceptable.

The calculation of brake horsepower depends upon the choice of type of compressor and number of stages. The brake horsepower per stage for the reciprocating compressor can be determined as (GPSA, 2004):

$$BHP = 0.0854 \left(\frac{Z_s + Z_d}{2} \right) \left[\frac{Q_{air} \times T_s}{E \times \eta_C} \right] \left(\frac{K}{K-1} \right) \left[\left(\frac{P_d}{P_s} \right)^{\frac{K}{K-1}} - 1 \right] \qquad (5)$$

where:

BHP: brake horsepower per stage
Q_{air} : air volumetric flow, MMSCFD
T_s: suction temperature, °R
P_1, P_2: pressure at suction and discharge flanges, respectively, psia
η_C : compressor efficiency
E : parasitic efficiency
High-speed reciprocating units – use 0.72 to 0.82
Low-speed reciprocating units – use 0.72 to 0.85

In the equation above, the parasitic efficiency (E) accounts for mechanical losses, and the pressure losses incurred in the valves of reciprocating compressors (the lower efficiencies are usually associated with low pressure ratio applications typical for pipeline compression). Hence, suction and discharge pressures may have to be adjusted for the pressure losses incurred in the pulsation dampeners for reciprocation compressors. The compression efficiency accounts for the actual compression process.

Finally, the maximum pressure and the temperature that the compressor can achieve are:

$$P_{max} = P_s \left[\frac{Z_d}{Z_s C} \left(1 - L - VE + C \right) \right]^K \qquad (6)$$

$$T_d = \frac{T_s}{\eta} \left[\left(\frac{P_d}{P_s} \right)^{\frac{K}{K-1}} - 1 \right] + T_s \qquad (7)$$

Note, if the discharge temperature is too high (more than 300 °F), additional cooling of the suction gas is required. It is recommended that the compressors be sized so that the discharge temperatures for all stages of compression be below 300 °F (Arnold and Stewart, 1999).

Cooler

For the cooler, the enthalpy or heat flow of the energy stream is removed from the cooler process side holdup:

$$F \times (H_{in} - H_{out}) - Q_{cooler} = \rho \frac{d}{dt}(V \times H_{out})$$ (8)

Where, F and H are the air flowrate and enthalpy, respectively; Q, duty, and V is the tube holdup. The duty applied to cooler can be calculated using exist stream's temperature specification.

The pressure drop of the cooler can also be determined in one of two ways; specify the pressure drop manually, and/or define a pressure flow relation in the cooler by specifying a k-value, which is used to relate the frictional pressure drop and flow through the cooler. This relation is similar to the general valve equation:

$$F = \sqrt{\rho} \times k \times \sqrt{P_{in} - P_{out}}$$ (9)

This equation uses the pressure drop across the cooler (without any static head contributions) to size the cooler with a k-value.

Air Receiver

For the air receiver, the mass balance is as follows:

$$\frac{d(\rho_{out} V)}{dt} = F_{in}\rho_{in} - F_{out}\rho_{out}$$ (10)

Where, F and ρ are the flowrate and the density of the air, respectively; and V is the volume of the air in the receiver. Subscripts 'in' and 'out' denote the inlet and outlet streams, respectively.

The energy balance is also as follows:

$$\frac{d}{dt}\left[(e+k+\varphi)V\right] = F_{in}\rho_{in}\left(e_{in}+k_{in}+\varphi_{in}\right) - F_{out}\rho_{out}\left(e_{out}+k_{out}+\varphi_{out}\right) \quad (11)$$
$$+ Q - \left(F_{out}P_{out}-F_{in}P_{in}\right)$$

Where, e, k, and φ are the internal energy, kinetic energy, and potential energy of the air (per unit mass), respectively; V, Q and P are the volume of the air, the heat lost across the boundary, and the stream pressure, respectively.

Air Valves

For the air flow through the ball valve, the resistance equation is written as follows:

$$\dot{m}_{air} = 29.68 \times C_v \times \sqrt{\rho_{air} \times P_{in}} \times \sin\left(122.035\sqrt{1-\frac{P_{out}}{P_{in}}}\right) \quad (12)$$

Where, $\dot{m}_{air}$ and ρ_{air} are the air mass flowrate (lb_m/hr), and density (lb_m/ft^3), P_{in} and P_{out} are the pressure of the inlet stream, and pressure of the exit stream without static head contributions (psi), respectively; and C_v is the valve capacity coefficient.

The value of C_V can be supplied by vendors. It is dependent on valve position and also varies with the valve. The mathematical relationship of C_V (%) and valve position (%) for equal percentage type of valves is as follows:

$$\% C_v = \left(\% \text{ Valve Opening}\right)^3 \quad (13)$$

Note, HYSYS reports the full C_V (at 100% open, which remains fixed) plus the valve opening. If the valve is 100% open, then we get a smaller valve than if the valve was only 50% open for the same conditions. This is just one way of sizing a valve as some sources report an effective C_V (varies with the valve opening) versus the valve opening.

OIL/WATER SUPPLY LINES

Pumps

Calculations are based on the standard pump equation for power, which uses the pressure rise, the liquid flowrate, and density:

$$Power = \frac{(P_{out} - P_{in}) \times Q_L}{\rho_L \times \eta_P} \tag{14}$$

Where P_{in} and P_{out} are the pump inlet and outlet pressures, respectively, Q_L, and ρ_L are the liquid flowrate and the liquid density, respectively, and η_P is the pump efficiency. When the efficiency is less than 100%, the excess energy goes into raising the temperature of the outlet stream.

The actual power is also equal to the difference in heat flow between the outlet and inlet streams:

$$Power = \left(Heat\ Flow_{out} - Heat\ Flow_{in} \right) \tag{15}$$

If the feed is fully defined, only two variables of the outlet pressure, efficiency, and pump energy need to be specified for the pump to calculate all unknowns. HYSYS can also back-calculate the inlet pressure.

Liquid Valves

For the liquid flow through the valve, the resistance equation is as follows:

$$\dot{m}_L = 63.338 \times C_v \times \sqrt{\rho_L} \times \sqrt{P_{in} - P_{out}} \tag{16}$$

Where, $\dot{m}_L$ is the liquid mass flowrate (lb_m/hr); ρ_L, density of the liquid (lb_m/ft^3), and P_{in}, and P_{out} are the pressure of the inlet stream, and pressure of the exit stream without static head contributions (psi), respectively.

Mixer

A complete material and heat balances is performed with the mixer. That is, the one unknown temperature among the inlet and outlet streams is always calculated rigorously. If the properties of all the inlet streams to the mixer are known (temperature, pressure, and composition), the properties of the outlet stream is calculated automatically since the composition, pressure, and enthalpy is known for that stream. Note, the resultant temperature of the mixed streams may be quite different than those of the feed streams due to mixing effect.

Topside/Three-Phase Separators

The physical parameters associated with these operations are the pressure drop across the vessel and the vessel volume. The pressure drop is defined as:

$$P = P_V = P_L = P_{Feed} - \Delta P \tag{17}$$

where:

P: vessel pressure
P_V: pressure of vapor product stream
P_L: pressure of product liquid stream(s)
P_{Feed}: pressure of feed stream (assumed to be the lowest pressure of all the feed streams).
ΔP : pressure drop in the vessel (the default pressure drop across the vessel is zero).

The vessel volume, together with the setpoint for liquid level/flow, defines the amount of holdup in the vessel. The amount of liquid volume, holdup, in the vessel at any time is given by the following expression:

$$H = V \times \frac{LL}{100} \tag{18}$$

Where, H is the holdup; V, vessel volume, and LL is the liquid level in the vessel at any time.

The vessel pressure is a function of the vessel volume and the stream conditions of the feed, product, and the holdup. The pressure in the holdup is calculated using a volume balance equation. Holdup pressures are calculated simultaneously across the flowsheet. The vessel level can also be calculated from the vessel geometry, the molar holdup, and the density for each liquid phase. The basic equations used for sizing two and three-phase separators in HYSYS can also be found in the papers published by Svrcek and Monnery (1993), and Monnery and Svrcek (1994).

REFERENCE

Arnold, K., and Stewart, M., "Surface Production Operations, Vol. 2: Design of Gas-Handling Systems and Facilities," 2nd Edition, Gulf Professional Publishing, Houston, TX (1999).

GPSA Engineering Data Book, 12th Edition, Gas Processors Suppliers Association, Tulsa, OK (2004).

Hyprotech Ltd., HYSYS.Plant Release 2.4, User Guide, Calgary, Canada (2001).

Monnery, W.D., and Svrcek, W.Y., "Successfully Specify Three-Phase Separators", *Chemical Engineering Progress,* 218-229 (1994).

Svrcek, W.Y., and Monnery, W.D., "Design Two-Phase Separators Within the Right Limits", *Chemical Engineering Progress,* 53-60 (Oct. 1993).

APPENDIX B

OLGA TRANSIENT CODE

OLGA is a commercial multiphase pipeline simulator which has been developed in 1983 by IFE and SINTEF for offshore applications. The physical model of OLGA was originally based on an extended two-fluid model, details of which are given below. The latest version, OLGA2000, includes a three-phase model, however, as there are few published details of this model, this discussion will center on the two-phase formulation.

OLGA's equation set was described by Bendiksen et al. (1991) and differs slightly from the standard six-equation model due to inclusion of a droplet field. This is a development to assist in the modelling of annular flow within a standard two-fluid model. The six equations to be solved are those for conservation of mass in the gas core, the liquid film and the droplet field, conservation of momentum in the liquid film and the core, and total energy conservation.

The continuity equations are:

$$\frac{\partial}{\partial t}\left(\varepsilon_G\, \rho_G\right) = -\frac{1}{A}\frac{\partial}{\partial x}\left(A\,\varepsilon_G\, \rho_G\, U_G\right) + \Gamma_G \tag{1}$$

$$\frac{\partial}{\partial t}\left(\varepsilon_L\, \rho_L\right) = -\frac{1}{A}\frac{\partial}{\partial x}\left(A\,\varepsilon_L\, \rho_L\, U_L\right) - \frac{\varepsilon_L}{\varepsilon_L + \varepsilon_d}\Gamma_G - \Gamma_e + \Gamma_d \tag{2}$$

and:

$$\frac{\partial}{\partial t}(\varepsilon_d\,\rho_L) = -\frac{1}{A}\frac{\partial}{\partial x}(A\,\varepsilon_d\,\rho_L\,U_d) - \frac{\varepsilon_d}{\varepsilon_L + \varepsilon_d}\Gamma_G + \Gamma_e - \Gamma_d \qquad (3)$$

The OLGA model treats the droplets in the gas as a separate phase. This additional droplet phase is used to account for additional longitudinal pressure drop in separated flow regimes such as annular and stratified flows.

When formulating the momentum equations, the gas and liquid droplet phases are combined into a single relation (Equation 4), where this technique eliminates the necessity for modeling the drag force on the liquid particles.

$$\frac{\partial}{\partial t}(\varepsilon_G\,\rho_G\,U_G + \varepsilon_d\,\rho_L\,U_d) + \frac{1}{A}\frac{\partial}{\partial t}(A\,\varepsilon_G\,\rho_G\,U_G^2 + A\,\varepsilon_d\,\rho_L\,U_L^2) = -(\varepsilon_G + \varepsilon_d)\frac{\partial P}{\partial x} -$$

$$\frac{1}{2}f_G\,\rho_G\,U_G|U_G|\frac{S_G}{A} - \frac{1}{2}f_i\,\rho_G\,U_r|U_r|\frac{S_i}{A} + (\varepsilon_G\,\rho_G + \varepsilon_d\,\rho_i)g\cos\theta + \qquad (4)$$

$$\Gamma_G\frac{\varepsilon_L}{\varepsilon_L + \varepsilon_d}U_a + \Gamma_e U_i - \Gamma_d U_{dep}$$

The liquid momentum equation is:

$$\frac{\partial}{\partial t}(\varepsilon_L\,\rho_L\,U_L) + \frac{1}{A}\frac{\partial}{\partial t}(A\,\varepsilon_L\,\rho_L\,U_L^2) + \varepsilon_L\frac{\partial P}{\partial x} = -\frac{1}{2}f_L\,\rho_L\,U_L|U_L|\frac{S_L}{A} +$$

$$\frac{1}{2}f_i\,\rho_G\,U_r|U_r|\frac{S_i}{A} + (\varepsilon_L\,\rho_L)g\cos\theta - \Gamma_G\frac{\varepsilon_L}{\varepsilon_L + \varepsilon_d}U_a - \Gamma_e U_i + \Gamma_d U_{dep} - \qquad (5)$$

$$\varepsilon_L\,D(\rho_L - \rho_G)g\frac{\partial U_L}{\partial x}\sin\theta$$

In Equations 1 through 5, $\varepsilon_d, \varepsilon_L$, and ε_G are the holdups in the droplet core, liquid film and of the gas, respectively; Γ_e, Γ_d, and Γ_G are mass transfer terms referring to entrainment, deposition and gas-liquid respectively; A is the pipe cross-sectional area, ρ, U, and P are the density, velocity, and pressure of each phase respectively; U_i, U_{dep}, U_r, U_a are the interfacial velocity, the deposition velocity, relative velocity between the gas and liquid (slip velocity), and the velocity of slug control volume boundary (which is used to account for the momentum source term) respectively; θ is the pipe inclination with the vertical, f is a friction coefficient and S is the wetted perimeter (subscripts G, L, and i indicate gas, liquid, and interface, respectively).

The friction factors and wetted perimeters are dependent on flow regime. The transition between the distributed and separated flow regime classes is based on the assumption of continuous average void fraction and is determined according to a minimum-slip concept. That is, the flow regime yielding the minimum gas velocity is chosen. Wallis (1970) empirically found a similar criterion to describe the transition from annular to slug flow very well. This criterion covers the transition from stratified to bubble flow, stratified to slug flow, annular to slug flow, and annular to bubble flow.

The above set of conservation equations is written in a general form in order to apply for all flow regimes. Observe, however, that certain terms may drop out for certain flow regimes; e.g. inside slugs all droplet terms disappear (Straume et al., 1992).

In the OLGA model, a pressure equation is obtained by adding the mass equations expanded with respect to pressure, temperature, and composition, neglecting variations of density with composition. The pressure and momentum equations enable a direct solution of pressure and average phase velocities, as described by Bendiksen et al. (1991).

To close the equation system described above, initial and boundary conditions, closure laws for friction in the two momentum equations as well as relations for mass transfer between the phases, are introduced, where the details are given by Bendiksen et al. (1991).

Finally, the mixture energy-conservation equation is:

$$\frac{\partial}{\partial t}\left[\sum_j m_j \Theta_j\right] + \frac{\partial}{\partial t}\left[\sum_j m_j U_j E_j\right] = H_S + Q_W \qquad (6)$$

Where, subscript j refers to each phase gas, liquid, and droplet. Q is the heat transfer; H_S is the enthalpy associated with the mass source, and Θ_j and E_j are given by:

$$\Theta_j = e_j + \frac{1}{2}U_j^2 + gh \qquad (7)$$

$$E_j = H_j + \frac{1}{2}U_j^2 + gh \qquad (8)$$

In the above equations, e denotes the internal energy of the fluid phase per unit mass; H is the enthalpy, U is the heat transfer from pipe walls, and h is the elevation.

The physical model of OLGA, as formulated previously, yields a set of coupled first-order, nonlinear, one-dimensional partial differential equations with rather complex coefficient. This nonlinearity means that no single numerical method is optimal from all points of view. OLGA solves the equations numerically using a semi-implicit technique in three steps. First, the pressure equation is solved with two conservation of momentum equations using a Gauss inversion technique based around the banded structure of the resultant matrix. Next, the three continuity equations are dealt with. Finally, the energy equation is solved. This system is however over-specified and so at each time step there is a small error in mass conservation. This mass leakage term is, therefore, calculated and incorporated into the pressure equation as a correction term at the next time step. Details can be found in a paper by Bendiksen et al (1988).

In the literature, a substantial number of transient flow case studies have been reported using OLGA software, capable of simulating normal offshore pipeline problems (severe slugging, pigging, etc.). However, Burke et al. (1993) highlighted one of the shortfalls of the OLGA model, where the slug frequency within the pipe needs to be specified. Since this parameter is not modeled adequately, this imposes a significant limitation on the application of the slug model within OLGA. This is most limiting in the design of new pipelines where there is no experimental data available to tune the simulation.

REFERENCES

Bendiksen, K.H., Espedal, M., and Malnes, D., "Physical and Numerical Simulation of Dynamic Two-Phase Flow in Pipelines with Application to Existing Oil-Gas Field Lines", paper presented at the Conference on Multiphase Flow in Industrial Plants, Bologna (September 27-29, 1988).

Bendiksen, K.H., Malnes, D., Moe, R., and Nuland, S., "The Dynamic Two-Fluid Model OLGA: Theory and Application", *SPE Production Engineering,* 171-180 (May 1991).

Burke, N.E., Kashou, S.F., and Hawker, P.C., "History Matching of a North Sea Flowline Startup", *Journal of Petroleum* (May 1993).

Straume, T., Nordsveen, M., and Bendiksen, K., "Numerical Simulation of Slugging in Pipelines", paper presented at the ASME International

Symposium on Multiphase Flow in Wells and Pipelines, 144, 103-112, *ASME-FED* (1992).

Wallis, G.B., "Annular Two-Phase Flow, Part I: A Simple Theory", *ASME Journal of Basic Engineering,* 92, 59-72 (1970).

APPENDIX C

DETERMINING PIPE SURFACE ROUGHNESS

Determining the exact value of the pipe surface is very important as it directly impacts the pressure drop of the pipeline-riser system. However, the values reported by pipe mills are just a guess, and normally do not reflect the exact average value of the roughness throughout the length of the pipeline (Mohitpour et al., 2003). Hence, the most accurate way of determining this value is during the operation, where we gather the real SCADA data for a long period of time, and then choose the steady state conditions for pressure and flowrates.

In order to estimate appropriate value of the pipe roughness, a single-phase water test was carried out into the pipeline-riser system, where processing the obtained results of the pressure drop over the riser extracted the desired pipe roughness. The results were processed as follows:

1. The frictional pressure drop over the riser (ΔP_f) is calculated as:

$$\Delta P_f = (P_B - P_S) - \rho_L \, g \, h_R \tag{1}$$

where:

P_B = riser-base pressure

P_S = topside separator pressure

ρ_L = liquid density

g = gravitational constant

h_R = riser height

Historically, pressure losses due to friction are evaluated using the friction factor and the Darcy-Weisbach equation. This equation which incorporates the Moody (1944) friction factor is:

$$\Delta P_f = \frac{f_M}{2}\left(\frac{L}{D}\right)\left(\frac{\rho_L U^2}{g_c}\right) \tag{2}$$

Where f_M is the Moody friction factor; L is the length of the pipe; D is the diameter of the pipe; U is the velocity of the fluid, and g_c is gravitational conversion factor. Note, sometimes, the Moody friction factor is replaced by Fanning friction factor (f_F) where:

$$f_M = 4f_F \tag{3}$$

Consequently great care must be taken when choosing the value of f with attention taken to the source of that value.

2. The friction factor can then be calculated from measuring the steady-state pressure drop across the riser section length using Equation 2.
3. Finally, the pipe surface roughness, ε, can be obtained using the von Karman equation for rough pipes (Fogarasi, 1975):

$$\left(\frac{1}{f_F}\right)^{0.5} = -4\log\left(\frac{\varepsilon}{3.7D}\right) \tag{4}$$

An estimated value of the pipe roughness is then used as input to the OLGA model of the pipeline-riser system, in order to appropriate prediction of the pressure drops and flow regimes into the system.

REFERENCES

Fogarasi, M., "Further on the Calculation of Friction Factor for Use in Flowing Gas Wells", *Journal of Canadian Petroleum Technology,* 14, 2, 53–54 (1975).

Mohitpour, M., Golshan, H., and Murray, A., "Pipeline Design and Construction, A Practical Approach", 2nd edition, *ASME Press*, ASME, New York, USA (2003).

Moody, L.F., "Friction Factors for Pipe Flow", *Trans. ASME,* 66, 671–684. (1944).

APPENDIX D

DETERMINING PIPELINE CRITICAL LIQUID VELOCITY AND LIQUID HOLDUP

This appendix details the calculations used to determine the pipeline critical liquid velocity and liquid holdup into the Boe (1981) stability criterion.

Assuming steady, fully developed stratified flow in the pipeline, where the gas and liquid flowrates are assumed to be low, and suggesting that the liquid holdup is approximately constant along the pipeline for all time; its value will be given by the geometrical relation as (Barnea and Taitel, 1986):

$$\varepsilon_{LP} = \frac{A_L\left(U_{SL}^0, \theta, D\right)}{A} = \frac{1}{2\pi}\left[\omega\left(U_{SL}^0, \theta, D\right) - \sin\omega\left(U_{SL}^0, \theta, D\right)\right] \qquad (1)$$

Where, ε_{LP} is the liquid holdup in the pipeline; A, pipe cross-section area; A_L, cross-section occupied by liquid, U_{SL}^0, liquid superficial velocity at the pipe inlet; θ, pipe inclination with respect to horizontal position; ω, wetted angle; and D is the pipeline internal diameter.

Solving the above equation for the wetted angle yields:

$$U_{SL}^0 - \frac{149}{2\pi}\left(\frac{D}{4}\right)^{\frac{2}{3}}|\sin\theta|^{\frac{1}{2}}\frac{(\omega - \sin\omega)^{\frac{5}{3}}}{\omega^{\frac{2}{3}}} = 0 \qquad (2)$$

Note, when ε_{LP} increases to one, ω tends to 2π, and U_{SL}^0 converges to the critical value:

$$\left(U_{SL}^0\right)_{critical} = 149\left(\frac{D}{4}\right)^{\frac{2}{3}}\left|\sin\theta\right|^{\frac{1}{2}} \tag{3}$$

Above this value, the flow pattern in the pipe is dispersed or nearly liquid. Therefore, the severe slugging phenomenon cannot exist (Schmidt et al., 1985).

A general solution of Equation 1 has been presented graphically by Taitel (1986). However, Yeung (1996) proposed a simple technique for calculations of the pipe liquid holdup as:

$$\varepsilon_{LP} = 0.099 + 0.7771X - 0.2913X^2 + 0.0525X^3 - 0.0036X^4 \tag{4}$$

where:

$$X = \log(\psi) \tag{5}$$

$$\psi = \frac{(\rho_L - \rho_G)g\sin\theta}{\left(\frac{dP}{dx}\right)_{SL}} \tag{6}$$

$$\left(\frac{dP}{dx}\right)_{SL} = \frac{f}{D}\left(\frac{\rho_L U_{SL}^2}{2}\right) \tag{7}$$

As can be seen from Equations 4-7, the basis of this method is the calculation of the pipeline liquid holdup based upon the liquid superficial velocity.

REFERENCES

Barnea, D., and Taitel, Y., "Encyclopedia of Fluid Mechanics", N.P. Chereminisoff, Editor, 3, *Gas-Liquid Flows,* 403-474 (1986).

Schmidt, Z., Doty, D.R., and Dutta-Roy, K., "Severe Slugging in Offshore Pipeline-Riser Systems", *SPE Journal,* 27-38 (Feb., 1985).

Taitel, Y., "Stability of Severe Slugging", *Int. J. Multiphase Flow,* 12, 2, 203-217 (1986).

Yeung, H., "Flexible Risers Severe Slugging", Progress Report Number 3, Project 4, Managed Program on Transient Multiphase Flows, Cranfield University, Bedfordshire, England (1996).

INDEX

A

absorption, xiii
absorption coefficient, xiii
accuracy, 34, 35, 37, 59
active feedback, 19, 20
air, 34, 35, 36, 46, 53, 59, 60, 61, 62, 65, 89, 94, 95, 97, 98
alternative, 4, 17, 18
amplitude, 4, 44
API, 5, 7
appendix, 113
application, 4, 17, 18, 19, 106
asphaltene, 5
assets, 45
assumptions, 49
atmospheric pressure, 37
availability, 44
avoidance, 5

B

behavior, 6, 14, 16, 35, 36, 37, 45, 48, 52, 54, 55, 62, 65, 67, 77, 88, 91
benefits, 4
blocks, 11, 45, 51
boiling, 53
boundary conditions, 37, 44, 53, 54, 55, 65, 89, 105

bubble, xiv, 10, 14, 15, 19, 23, 27, 28, 29, 30, 31, 32, 62, 70, 71, 73, 75, 77, 88, 105
bubbles, 29, 31, 70

C

calibration, 45, 88
capital cost, 21
capital expenditure, 1
carrier, 92
cast, 45
cavitation, 50
cell, 54
classes, 15, 105
classical, 24, 36, 88
closure, 105
coal, 47, 48, 50, 80
codes, xi, 37, 59
components, ix, 2, 50
composition, 48, 91, 92, 100, 105
compressibility, 26, 94
computation, 56, 59, 76, 79
conductance, 93
configuration, ix, 3, 4, 21, 23, 50
conservation, 91, 103, 105, 106
constraints, 3, 6
consumption, 18
continuity, 103, 106
control, x, 12, 17, 19, 20, 21, 22, 43, 44, 45, 46, 48, 49, 50, 51, 52, 80, 85, 104

convergence, 56
conversion, 110
cooling, 17, 96
correlation, 27, 32
correlations, 25
corrosion, 5
cost-effective, 3, 5, 17
costs, 3, 21
critical temperature, 53
critical value, 79, 114
cross-sectional, 104
cycling, 13, 15, 33, 35, 36, 37, 51, 67, 69

D

definition, 3
degrees of freedom, 19, 92
delivery, 70, 73
density, 53, 76, 77, 78, 89, 95, 97, 98, 99, 101,
    104, 105, 109
deposition, 5, 104
deviation, 51
differential equations, 48
differentiation, 35
discretization, 54
displacement, 95
distribution, 32
duration, 77

E

energy, 46, 91, 92, 96, 97, 99, 105, 106
environment, 45, 48, 51, 52, 91
environmental conditions, 2, 3, 5
equilibrium, 91
execution, 53
experimental condition, x, 36
exploitation, ix, 4

F

fatigue, 5
feedback, 19, 20, 21
film, 70, 73, 103

floating, ix, 2
flooding, 9, 44
flow rate, ix, xiii, 7, 10, 11, 17, 18, 19, 21, 26,
    35, 46, 48
fluctuations, ix, 7, 15, 54, 60, 65, 89
fluid, 5, 22, 36, 50, 69, 70, 77, 91, 103, 106,
    110
freedom, 19, 92
friction, 9, 94, 104, 105, 110

G

gas phase, ix, 6
gas separation, 1
GPA, 56, 57
gravitational constant, 109
gravity, xiii, 9, 47
growth, 2, 24

H

handling, 19
hanging, ix, 4, 15, 16, 34
heat, 91, 95, 96, 97, 99, 100, 105, 106
heat capacity, 95
heat loss, 91
heat transfer, 105, 106
height, 15, 24, 110
high pressure, 12
hydrate, 5, 18
hydro, 2
hydrocarbons, 2
hydrodynamic, 6, 17, 29
hydrostatic pressure, 11

I

identification, 16
implementation, 21, 85
inclusion, 103
industrial, 37
industry, 1, 12
initiation, 28
injection, 17, 18, 30, 33, 34

instabilities, 54
insulation, 91
integration, 56, 66, 89, 92
integrity, 43
interactions, x, 65
interface, 3, 11, 37, 46, 48, 71, 77, 104
interphase, 9
inversion, 106
investigations, 34, 89

K

kinetic energy, 97

L

laws, 105
leakage, 94, 106
lifetime, 5
likelihood, 23
limitation, 106
liquid film, 103, 104
liquid phase, 6, 101
liquids, 17, 19, 22, 46, 70
long period, 109
losses, 43, 95, 96, 110

M

magnetic, iv
management, 45
manifold, 3
manifolds, 3
market, 45
mass transfer, 104, 105
matrix, 92, 93, 106
measurement, 35
mixing, 100
modeling, 44, 46, 49, 50, 91, 104
models, 19, 23, 44, 46, 59, 91
mole, xiii
molecular weight, 26, 53
momentum, 46, 103, 104, 105, 106
motion, 2

movement, 11
multiphase flow, 2, 5, 6, 9, 13, 20, 22, 43, 45, 48, 52, 59

N

natural, 44
nitrogen, 53
nodes, 92, 93
normal, 19, 53, 55, 88, 106

O

observations, 19, 88
offshore, iv, ix, x, 1, 5, 7, 12, 43, 44, 46, 47, 52, 59, 103, 106
offshore oil, 1
oil, iv, ix, 1, 2, 4, 36, 45, 46, 50, 53, 89
oil production, 2
operator, 44
optimization, 2
orientation, 46
oscillation, 11, 14, 36
oscillations, 12, 16
oxygen, 53

P

parameter, 31, 32, 106
partial differential equations, 106
particles, 104
periodicity, 55
permit, 48
philosophy, 45
physical properties, 89
pipelines, iv, 1, 5, 12, 43, 44, 55, 106
plants, 45
platforms, ix, 1, 2, 22
poor, 9, 12
power, 98, 99
precipitation, 5
prediction, 23, 24, 27, 31, 33, 34, 62, 67, 71, 75, 78, 88, 89, 110
prevention, 12, 17

propagation, 11, 27, 71, 77
proportionality, 30
pseudo, 53
pumping, 95
pumps, 46, 50

Q

quasi-equilibrium, 19

R

range, 5, 14, 36, 51, 53, 89
real time, 44
reality, 37, 65
reception, 46, 48, 88
recovery, 12
relationship, 98
relationships, 14, 45, 91
remediation, 5
reserves, 12
reservoir, 3
reservoirs, 2
resistance, 49, 92, 93, 94, 98, 99
response time, 21
returns, 92
roughness, 53, 109, 110

S

safety, 12, 22
satellite, 6
scaling, 5
seabed, ix, 2, 4
selecting, 3
sensitivity, 59, 76, 77, 89
sensors, 36
separation, ix, 1, 7, 9, 12, 20, 36, 43, 47
series, 18, 35, 36, 46, 54, 88
severity, 17, 18, 26
shape, 4
short period, 80
signals, 21

simulation, 37, 43, 44, 45, 48, 50, 53, 54, 55,
    59, 61, 65, 67, 71, 76, 88, 89, 91, 92, 106
simulations, 20, 44, 45, 53, 54, 55, 61, 71, 77
software, 43, 44, 48, 50, 52, 55, 106
speed, 48, 56, 94, 95, 96
S-shaped, ix, 15, 31, 33, 35, 36, 37, 40, 41, 57
SSI, 88
stability, 13, 15, 16, 19, 23, 30, 33, 34, 35, 54,
    55, 79, 80, 89, 113
stabilization, 22
stabilize, 19, 20
stages, 10, 43, 56, 73, 95, 96
stainless steel, 53
steady state, 11, 23, 45, 48, 109
steel, 5, 53
strategies, 5, 17, 20, 52
streams, 46, 50, 92, 97, 99, 100
supply, 46, 50
suppression, 21, 22
surface roughness, 110
surging, 13
switching, 88
systems, iv, ix, 1, 2, 5, 6, 9, 10, 16, 17, 18, 20,
    21, 22, 43, 44, 45, 46, 54, 85

T

tanks, 46
technology, 1, 2, 22
temperature, 3, 6, 20, 48, 53, 93, 95, 96, 97,
    99, 100, 105
tension, 2, 4
Thermal Conductivity, 54
thermodynamic properties, 53
threshold, 50
time, xi, 10, 12, 13, 15, 28, 35, 44, 45, 48, 54,
    55, 56, 62, 63, 64, 67, 72, 73, 74, 75, 76,
    77, 80, 91, 93, 100, 106, 109, 113
total energy, 103
total product, 1
transfer, 18, 105, 106
transition, 14, 24, 35, 46, 48, 62, 88, 105
transmission, 5
transport, 2, 5, 6, 9, 17, 43, 44
transportation, 1

turbulent, 93

U

uncertainty, 43

V

validation, 79
values, 31, 32, 53, 54, 67, 88, 93, 109
vapor, 100
variables, 22, 45, 92, 93, 99
    6

variation, 9, 10, 76
velocity, 6, 12, 24, 25, 28, 29, 31, 32, 33, 35,
    65, 69, 79, 104, 105, 110, 113, 114
vessels, 2

W

water, ix, 1, 3, 4, 22, 34, 35, 36, 45, 46, 50,
    53, 59, 60, 61, 65, 109
wear, 12, 50
wells, 6, 44